BOTTOM-UP BEEKEEPING
Learning from debris on the hive floor

By Ray Baxter

Editorial support from Ann Chilcott

Pollen contributions by Christine Coulsting

Northern Bee Books

Bottom-up beekeeping

Copyright © Ray Baxter

All rights reserved. No part of this publication may be reproduced, stored in a retrieval system, transmitted in any form or by any means electronic, mechanical, including photocopying, recording or otherwise without prior consent of the copyright holders.

Published 2025 by Northern Bee Books,
Scout Bottom Farm,
Mytholmroyd,
West Yorkshire
HX7 5JS (UK)
Tel: 01422 882751
Fax: 01422 886157
www.northernbeebooks.co.uk

ISBN 978-1-912271-96-2

Design and artwork DM Design and Print

Acknowledgements

I would like to offer a thousand thanks to Ann Chilcott, without whom this book would not exist. Ann is an author and esteemed apiculturist who gave me the greatest gifts that an author can receive: an honest and comprehensive critique of work, while also providing the support needed for a new writer to find their own voice and put pen to paper. Ann encouraged me to contact Northern Bee Books 'You should write a book about bee debris, and they should publish it' Well I did, and here it is. Thank you, Ann. I owe you several jars of honey!

Christine Coulsting's contributions were also essential. As a Master Beekeeper, she also has an extensive knowledge of honey bees and expertise in pollen identification. Christine worked closely with Maragaret Anne Adams to produce the book 'Pollen Grains & Honeydew: A guide for identifying the plant sources in honey'. Sadly, Margaret passed away shortly after the publication of this book. I was very fortunate to work with Margaret. My class would try to identify pollen found in honey and we sent our samples to Margaret for verification while we worked remotely during lockdown, trying to provide the best learning experience possible. My students loved working with her, and we all learned so much. Thank you, Margaret, I wish I could send you a copy of my book.

I dedicate this book to Paula Baxter who has put up with my ways for more than forty years. This book is a small token of my heart felt appreciation for over four decades of love, support and cake. Paula is very much the driving force on our flower farm and a passionate wildlife advocate, as can be seen by her contributions about moths found in this book.

Many thanks to everyone who contributed to the Facebook and Instagram groups that were created to share and discuss ideas during the research stage. I've been extremely grateful to the many beekeeping and apicultural professionals and amateurs who have taken the time to respond to my calls for help and advice, many of whom are acknowledged throughout. Any omissions are firmly my own oversight.

I also wish to thank the many young beekeepers I have taught over the years, they have always given me a great deal to think about and their questioning of the established way of doing things is a constant prompt to look more and assume less. Those who have inspired are too numerous to name.

Finally, thanks to the bees who have given me the greatest gift of all.

Contents

Chapter 1   Introduction   3

1.1	What is this book about, and what it isn't	3
1.2	Bee debris - what is it, where does it come from and why does it matter?	5
1.3	Book structure	10
1.4	Lessons from a lockdown hobby	11
1.5	Teaching beekeeping in high schools	14
1.6	Getting ready for home study	17

Chapter 2   Methods used   20

2.1	Repeat photo point recording	20
2.2	Measuring the weight of debris	22
2.3	Counting parts from the debris	23
2.4	Microscopy to identify pollen and other things	27

Chapter 3   Data visualisation of bee debris   29

Figure 3.1	Debris weight	29
Figure 3.2	Debris heat maps	30
Figure 3.3	Bee hairs	31
Figure 3.4	Chalkbrood cysts	32
Figure 3.5	Varroa mites	33
Figure 3.6	Minimum, maximum and mean temperatures in The Scottish Borders	33

Chapter 4 Debris tells the tale   34

4.1	January	Nearly a gold dance	34
4.2	February	Mites and textiles in the debris	48
4.3	March	Spring growth	61
4.4	April	Rapid growth	72
4.5	May	Moths and microplastics	85
4.6	June	A brood gap in June	101
4.7	July	Changing debris and changing methods	108
4.8	August	Drone culling	127
4.9	September	Shorter day length	142
4.10	October	Removing or adding stores	153
4.11	November	More fungi	161
4.12	December	Return of the mites	170

Chapter 4 Conclusion   178

Bibliography   187

Figures

1.1	Hive floor without an inspection board after 365 days
1.2	Six months of trapped debris under a ventilation screen
1.3	The first micro observation of the bee by Francesco Settuti
1.4	Reproductive organs of the honey bee by Jan Swammerdam
1.5	A hoard of early silver hammered coins
1.6	Early bee house built by Edward Bevan
1.7	Students studying bee debris in 2020
1.8	The bee house at Mill Pond Flower Farm
2.1	Repeat photopoint monitoring
2.2	Weighing by quadrat and heat maps
2.3	Sample preparation stage 1
2.4	Sample preparation stage 2
2.5	Sample preparation stage 3
2.6	Sample preparation stage 4
2.7	Evaporate the solvent
2.8	Stain the sample
2.9	Microscopy
3.3.1	Debris weight
3.3.2	Debris heat maps
3.3.3	Bee hairs
3.3.4	Chalkbrood cysts
3.3.5	Varroa mites
3.3.6	Minimum, maximum and mean temperatures in The Scottish Borders
4.1.1	January debris
4.1.2	Pollen found in January debris
4.1.3	A golden egg
4.1.4	A pair of eggs
4.1.5	Fibres found in the debris
4.1.6	Chalkbrood structures and Alternaria spp.
4.1.7	A lump of chalkbrood cysts found in the waste
4.1.8	Bee hairs
4.1.9	Spongy feet and grabbing claws
4.2.1	February debris
4.2.2	Fresh forage in february
4.2.3	February pollen
4.2.4	Hard body parts
4.2.5	Forehead of a bee
4.2.6	Bee parts - entombed fragments
4.2.7	Fibres found in the debris
4.2.8	Fibre attached to an amputated claw

4.2.9	Debris mites
4.2.10	Eggs in the debris
4.2.11	An emerging mite
4.3.1	March debris
4.3.2	Compressed pollen lumps
4.3.3	March pollen
4.3.4	Wax scales and stained chewed wax
4.3.5	Chewed virgin wax
4.3.6	Wax welded at the mid rib
4.3.7	Healthy varroa v damaged mite
4.3.8	Alternaria fungi
4.4.1	April hive debris
4.4.2	Brood frames from above the frames
4.4.3	The outer wax and propolis layer of a brood cap
4.4.4	The inner silk layer of a brood cap
4.4.5	Cell cappings by Frank Cheshire
4.4.6	Cocoon silk fibres by microscope
4.4.7	Chalkbrood in the debris
4.4.8	April pollen
4.4.9	A stellate trichome
4.4.10	Propolis entombing fungal cysts
4.4.11	Propolis entombing a costa
4.5.1	May debris
4.5.2	Eggs, young larvae and polished cells
4.5.3	Different coloured debris
4.5.4	An individual chalkbrood cyst
4.5.5	Black fungal structures
4.5.6	May pollen
4.5.7	Moths at the apiary
4.5.8	Fibres found in five Scottish hives
4.5.9	Fibres found in the debris by month
4.6.1	June debris
4.6.2	June pollen
4.6.3	Field beans in flower
4.6.4	Drawing comb
4.6.5	Fresh comb above the inspection screen
4.7.1	July debris - sampled weekly
4.7.2	Debris change in one week
4.7.3	Week 1 - propolis deposition
4.7.4	Week 4 - propolis deposition
4.7.5	July pollen

4.7.6   Debris after two days with a virgin queen
4.7.7   Twenty eight days of debris
4.7.8   Debris post-mating and pre-emergence
4.7.9   Debris produced by a virgin colony with a mated colony
4.7.10 Pollen found in week 1
4.7.11 Pollen found in week 4
4.7.12 Fungal mycelium in the debris
4.7.13 An emerging bee
4.7.14 Debris post-emergence
4.8.1   August debris
4.8.2   Cannibalised pupal parts collected from the debris
4.8.3   Cuticle parts
4.8.4   Typical antennae found in the debris
4.8.5   Bee bite marks
4.8.6   Varroa bite marks found only by microscopy
4.8.7   Hairy finds
4.8.8   Propolis under the microscope
4.8.9   Propolis samples dissolved in alcohol
4.8.10 Bee parts in dry propolis
4.8.11 Bee parts in dissolved propolis
4.8.12 August pollen
4.8.13 A summer honey crop of 37.8kg
4.9.1   September debris
4.9.2   Day length by month
4.9.3   Wasp model for hive inspections
4.9.4   Accessing food-stores?
4.9.5   Composition of food store debris?
4.9.6   Varroa visible by eye
4.9.7   Varroa parts shown by microscopy
4.9.8   September pollen
4.10.1 October debris
4.10.2 Whitish debris
4.10.3 Sugar crystals
4.10.4 Whole varroa
4.10.5 Alternaria species
4.10.6 Trichomes surrounded by ivy pollen
4.10.7 October pollen
4.11.1 November debris
4.11.2 The temperature on sampling day
4.11.3 Overwintering just fine
4.11.4 Field bean seedlings

4.11.5   November Alternaria
4.11.6   November pollen
4.12.1   December debris
4.12.2   Heat map for November and December
4.12.3   Debris away from the core
4.12.4   A concentrated sugar solution
4.12.5   Mites
4.12.6   Mite debris
4.12.7   Bee body hairs
4.12.8   December pollen

Bottom-Up Beekeeping *Ray Baxter*

Chapter 1. Introduction

1.1 What this book is about and what it isn't

This book is a guide through the world of debris produced by one honey bee colony, written for beekeepers and enthusiasts who want to better understand the lives of their bees. It explores using debris as a tool, alongside more traditional inspections, to monitor and manage bee colonies. Using my own experience, including lots of mistakes, this observational study will give you a place to start in understanding the debris produced by a colony, so that you can begin looking, record what you find, and use debris to answer some of your own questions about honey bee colonies and beekeeping. I hope that this information will be of practical interest to the beekeeper, DIY researchers, and anyone who marvels at the natural world.

I have no doubt that the process of writing this book has transformed my beekeeping practice and will continue to do so. So much so that before opening a hive, and certainly before moving any colony to my home apiary, the first thing I now do is look at the debris. It gives me the clues that I need to make judgments about the location of the colony, its size, the shape of the cluster, the history of the colony, colony activity, pests and parasites, and helps me to make informed hive interventions. Sitting here in my cosy bee house, with the gentle hum of bees right next to me, I can hand on heart say that I now have a deeper and broader understanding of my colonies than before. I hope to convince everyone who reads this book that each fragment of bee debris can tell a story and that learning to read the debris is an important skill for beekeepers. Also, recording how bee debris changes over time (more about this later) may be very important work in monitoring the status of bee health and the wider environment.

I have learned through careful observation of the debris that we can spot subtle changes, which can help to explain events from within the colony and the wider environment. I have spent countless hours going through scientific papers to better understand observations and trends. Sometimes this has been illuminating, often the search comes up blank. Frequently, it has felt like I am trying to tell a story that sits outside of scientific understanding. A story which needs more detective work to unravel its secrets.

This project is a first step towards the book that I wanted and needed twelve years ago, when I first started to keep bees. It is not meant to be a scientific paper, or a technical guide; it's more of a month-by-month diary with pictorial illustrations and commentary. Much more research over more than one season, and involving more colonies would be required to produce a definitive guide.

Whilst this account focuses on debris created by bees at my home colony in the Scottish Borders, the tips within it will hopefully be helpful to people in other localities. The debris profile of different colonies will have many similarities and differences, and these will be influenced by many variables, so much further study would be needed to make more confident conclusions. It is perhaps surprising that this work has not already been done, especially considering that collecting and analysing bee debris is so much easier and less invasive than many methods commonly used in bee research. Looking at bee debris is less disruptive for the bees than ripping apart a nest to find out more about the goings on within the colony.

In the research stage of this project a private discussion group was set up for people interested in the topic. Regular progress updates and photos of debris created lots of discussion and prompted further questions. I am extremely grateful to everyone who contributed to those discussions. Certainly, the outcomes are much better for it, but any mistakes remain firmly mine. This group became essential, providing support and challenge and becoming a kind of informal review process. The sharing of pictures and ideas undoubtedly helped me to better understand the topic, which in turn helped to develop more informed questions. Some of these questions are explored in the monthly diary and others are discussed further in the conclusion. Over time, monthly finds became the guide for these lines of enquiry. For example, in January microfibres and honey bee queen eggs were found, so these were investigated further, and in February mites were found so these were explored, and so on.

I was delighted when others in the group started their own debris research. They are interested and fascinated by the variety of things that are found in waste, such as living mites, pollen, fungi, recycled wax, cappings, cocoon silk, fibres, cannibalised bee parts, etc. The debris profile has a story to tell about the goings-on in a colony and the wider environment. People have even sent me their own photos of bee waste that they have come across and asked for an explanation. I am absolutely no expert, and they all know that, but I now know a bit more about colony debris than most people do, I love bees and I am happy to help. This book aims to encourage and inspire people to look a bit more, to learn more about the bee debris and what it can tell you about honey bees and the wider environment. I am writing this book so that, when you ask your questions, I can direct you to a source for answers, and you have a place from where to start your own debris journey.

1.2 Bee debris – what is it, where does it come from and why does it matter?

Imagine that you are a bee standing on the inspection board, looking up at the 10 frames hanging above you. Standing on your back legs, you are around 12mm tall. I'm almost 150 times taller than you at just over 1800mm. Each frame is 21 times taller than you. For me that is the same as a 14-storey building. If we add two more brood boxes, climbing to the top would be higher than scaling Big Ben. I wouldn't recommend trying to run up the 334-stepped spiral stairway to the belfry where Big Ben, the great bell, hangs. It's a task that appears effortless for bees. Now imagine you are peering up at the 5000 multi-use apartments on either side of the 10 buildings (frames) above you. In total, over 100,000 mini apartments or cells in the 10 high-rise frames in this building complex. These skyscraper apartments are used in different ways throughout the year. In the winter, many apartments are empty, but in the summer bees and beekeepers like them to be fully occupied.

Humans produce an enormous amount of waste, with the average UK citizen creating 1kg of waste daily, or about 40 tons per day for 40,000 people. Imagine now that a fraction of that 40 tons falls between the buildings onto the streets below. A rapid team of refuse collectors would be needed to clean up the mess and make the space safe, sanitary and liveable. This waste burden has a significant cost for human populations who today are grappling with how to improve standards for reusing and recycling the waste of consumption.

In a hive, this clean up exercise is complicated by the human intervention of adding a ventilation screen. The fine mesh screen allows some waste to fall through and prevents bees access to reuse, recycle or dispose of the material that falls through this screen. The ventilation screen creates a debris trap, for a mixture of waste and valuable resources that could be re-used by the bees. Introducing such a ventilation screen on our streets would be a health and safety nightmare for human populations. Thankfully, the inspection board used by beekeepers can be easily removed and cleaned. The debris analysed in this study is simply any material that has fallen through the ventilation screen and lands on the inspection board. It would be wrong to describe this material as waste, because if the bees could access it then much of it would be reused, recycled or otherwise disposed of. To illustrate, a solid floor without a ventilation screen was added below a full double brood box of bees in 2023. Figure 1.1 shows the debris found 365 days later. In one year, no debris was visible on the solid floor, apart from a lump of propolis at the entrance (more about this later). The lack of debris on this board hints at the remarkable waste management system employed by bee colonies. No doubt human populations could learn a great deal about waste management from studying how bees reuse, recycle and dispose of waste. Contrast this 'clean' hive floor with figure 1.2 which shows 6 months of debris trapped under a ventilation screen. I stumbled upon this abandoned hive in a damp woodland. While looking at this collection of debris, a few workers came to greet me showing that the colony was still alive despite the additional stress factor of decomposing debris.

Figure 1.1 Hive floor without an inspection screen after 365 days

Imagine you are a bee

While this game of "imagine you are a bee" can be fun and useful, it is a risky business that can sometimes get us into trouble when trying to understand the colony. As humans, the entirety of our learning experience and worldview is informed by our sensory mechanisms and by the evolutionary knowledge contained in our DNA. The way we see, touch, feel, smell, hear and taste are the gateways to everything that we understand about the world. It is impossible to imagine how you would learn or perceive the world with the sensory alien superpowers of a bee. Everything that we understand about our sensory world would be different. The way we touch, see, hear, taste, perceive time, space and electro-magnetic fields would be profoundly different. How can we possibly imagine such a world when there are no equivalent sensory systems in humans? The risk is anthropomorphism on a grand scale and relying on human models such as The British Monarchy to describe the honey bee colony. Such a simple story robs us of the real poetry of the hive. This poetry is the wonderous reality that is hard earned, through the rigour of scientific endeavour, with each discovery revealing a little more honey bee knowledge. Perhaps one day, Artificial Intelligence will be able to mimic the bee brain and analyse sensory information from human-made models of bee sense organs. Perhaps this sensory information will offer new ways to problem solve and produce creative outputs. A regret that I have about getting older (still in denial) is not living long enough to witness the revolution in intelligence that is upon us. Perhaps in the decades to come people will be better equipped to imagine being a bee.

Figure 1.2 Six months of debris trapped below the ventilation screen

The process of writing this book has been to make lots of observations of the debris and ask questions later. It's an approach that would not score well in high schools, which insist that scientific endeavour starts with a good question or hypothesis. Of course, as a science teacher I am a passionate advocate of the importance of hypothesis driven scientific method, but something is missing in a science classroom that excludes open ended detailed observational study. Also, as discussed earlier, if we are unable to imagine being a bee, how is it then possible to know what questions to ask about the life of a honey bee. Collecting lots of data and then asking questions later seems like a good way forward. I expect that by the end of this observational study I will be able to ask much more informed questions about the debris created by honey bee colonies. A hypothesis-driven study may well come later.

The early pioneers of microscopy would be able to tell us what is lost from the school curriculum. Take for example the anatomical details of the bee shown in figure 1.3 (Stelluti, 1630) or figure 1.4 showing the work of Swammerdam who in the seventeenth century was the first to use microscopy to describe the ovaries and the male reproductive system of bees (Dodd, 2000).

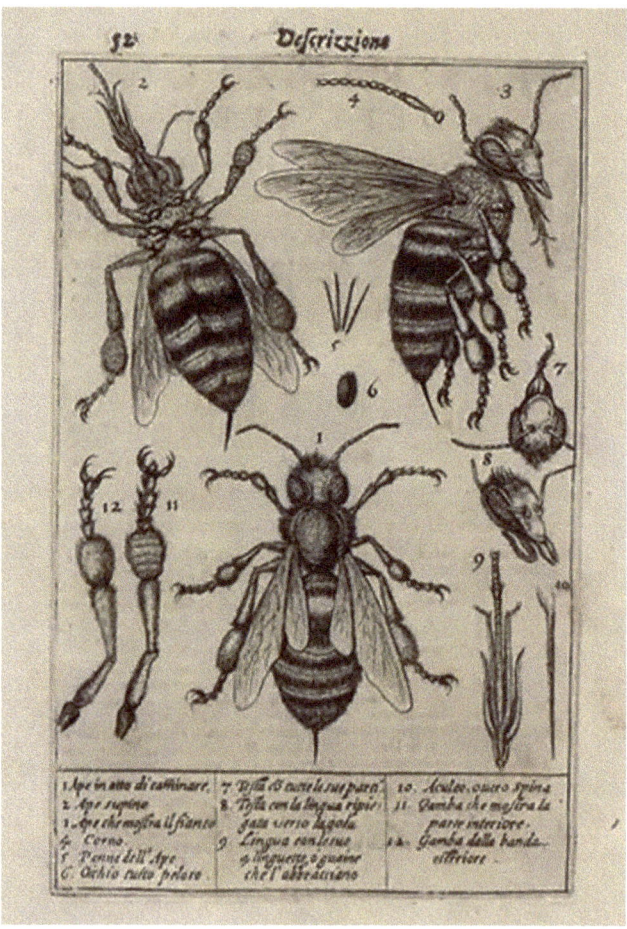

Figure 1.3 The first micro-observation of the bee by Francesco Stelluti

Swammerdam's detailed observations confirmed the earlier observations of (Butler, 1623) who was one of the first to assert in a radical publication titled 'The Feminine Monarchie' that said that there was in fact no king bee, which was the common belief at the time. We can perhaps imagine the joy of Charles Butler, Jan Swammerdam or Stelluti in making these observations. There is a joy to be gained through mindful observation, shared by everyone who finds something new, identifies it and learns something that they did not know before. My wife is a passionate moth recorder and I see the same joy in her eyes when she shares her moth records.

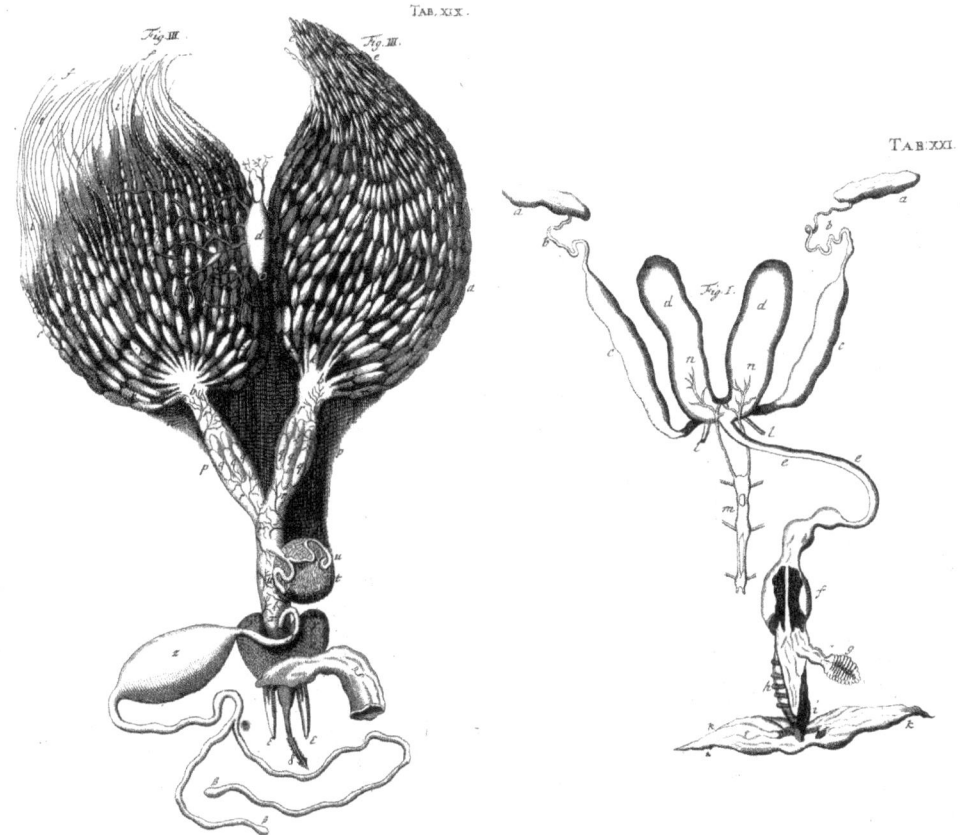

Figure 1.4 Reproductive organs of the honey bee by Jan Swammerdam

I have made more mistakes than most as a beekeeper. In the beginning, I got carried away with hive inspections, trying to learn about the colony. Understanding what's going on in the hive is important work for all beekeepers. However, over-inspecting can be a real problem. Inspections are stressful for bees and people, disturb the conditions in the hive and often don't help to create a beekeeping plan. Of course, there is a time and place for inspections, as there is also a time to leave well alone. Learning when and how to intervene are essential skills for any beekeeper.

A frequent mistake during a hive inspection is a failure to prepare properly. I have learned the hard way that patience, being prepared and doing your research are the most important attributes for a beekeeper. I now consider learning how to read debris as an essential part of doing my research and planning to make positive hive interventions.

1.3 Book Structure

Chapter 2 includes a summary description of the main methods used in this study. This is meant to be a simple guide for anyone interested in carrying out a similar study or adapting the methods for a different purpose. Chapter 3 includes a visual representation of the result outputs created from the methods used. Anyone who likes data visualisation and graphs will probably want to turn straight to Chapter 3. You may however choose to skip this chapter and return to this section later as you read the main part of this book, when each graph will be discussed in a timely manner. This section is an annual diary of the debris produced by one honey bee colony (Chapter 4). This diary record became a prompt for further reading, for discussion and a tool to develop thinking, asking questions and finding answers.

As the months passed, this diary grew arms and legs, I felt the urge to go back and rewrite this chapter and "join the dots". A deliberate decision was made not to do that, rather to leave each month's observations to stand alone "warts and all". This diary is a real time record of what happened during my own learning journey. The final chapter uses the diary experience along with the results from this home study to discuss three main questions.

- How does bee debris help to better understand the colony and what do we need to find out?
- How does bee debris help to better understand the wider environment and questions for further study?
- Ways to use the study of debris for positive hive interventions

Finally, the bibliography lists the reports and publications used to describe and explain the debris collected. Most of these include a web-address for open-source publications with the full text available for everyone who wants to read further.

1.4 Lessons from a lockdown hobby

Obviously, beekeeping is my first favourite pastime, but pottery and metal detecting are close seconds. You may be surprised that there are similarities between this DIY research project and metal detecting. Metal detecting has changed my understanding of the local history near my home. In the same way, I am learning to better understand my bees by monitoring the inspection board. It's not glamorous work, but debris has a huge story to tell about the colony and the wider environment.

Studying ancient rubbish gives archaeologists and historians a way to better understand the behaviours that once defined people's daily lives. It may not be glamorous work, but much of what we understand about the ancient world comes from rubbish. Human rubbish is a product of human consumption and in turn tells a story about human behaviour, likewise, bee waste can tell a story about bee behaviour and the wider environment.

During the first lockdown, my wife gave me a metal detector saying it could be daily exercise. Of course, I knew that she wanted me out of the house during those crazy stay-at-home days. I was out in all weathers, researching local history, asking landowners for permissions and looking for 'treasure'. I am now quite expert at identifying ring pulls, buckles, bullets, buttons and all sorts of rubbish. Should you ever visit my home I will proudly give you a tour of my collection of thimbles or buttons through the ages. Equal to the excitement of finding stuff, is the process of identifying the 'treasure'. Sometimes it can take weeks of research to make a positive identification and sometimes it's not possible at all. From the start, I purposefully decided to detect only within walking distance of my home. Also, the sharing of the stories and history of these finds with local people creates genuine interest in better understanding the place in which we live.

It took over a year of daily detecting until I found anything exciting. Then one wet, grey Scottish afternoon in a field with no recorded history, I started to dig a wide range of medieval objects dating from the eighth to the thirteenth centuries. Imagine my joy to find a hoard of silver hammered coins from the reign of Edward I the King of England (see figure 1.5). To be the first person to hold these coins for 800 years; well, the feeling was a little spooky. My mind was teleported back to the 1200s.

Over the next few months, a picture began to emerge of a medieval village on the find site. I started to create a mental picture of life in that field, using the remnants of cooking pots, buckles, gold gilded medieval horse pendants, toy dogs, pencils made from lead, spinning whorls, and an appropriately named beehive thimble from the 1200s. There is such a buzz (no pun intended) in rescuing these objects from the earth. I swear locals thought I was bonkers, up to my knees in mud looking for bits of metal, but for me it was like time travel, and became the best way to understand local history.

Figure 1.5 A hoard of early silver hammered silver coins

Getting back to bees

Standing in this muddy field, with pockets full of 'treasure', I tried to picture the best place for an apiary, using barley straw skeps. It is possible that local skeps were made from barley, because this part of Berwickshire was famous for producing large quantities of barley. In fact, the very name Berwick means barley town.

It is possible that the people who once owned this medieval 'treasure' had an apiary nearby, perhaps one like the magnificent silver Standing Cup presented to the Worshipful Wax Chandlers in the 1600s. This ancient cup celebrates medieval beekeeping and depicts a large wooden framed apiary standing with three straw skeps. This beautiful inscription tells a story about beekeeping practice in a time before the discovery of removable frames and shows some of the many uses of beeswax. The engraving also shows a beekeeper 'tingling' a swarm of bees, someone melting wax in a cauldron above a fire and a figure lighting a candle beside a reading desk. Another glimpse into beekeeping practice of days gone by is illustrated by a wonderful beehouse built more than 200 years ago by Edward Bevan in Herefordshire (see figure 1.6)

"I cannot give a better notion of what I consider to be the most eligible plan for a beehouse than describe my own. The frame is built from timber and plastered on the inside. The bee building is not only thatched on top, but down the sides and ends. The floor is boarded. Windows on either side allow the light to fall sideways of the hive" (Bevan, 1838).

Figure 1.6 Early bee house built by Edward Bevan

Perhaps these mediaeval beekeepers could teach us all a thing or two about bee husbandry. Skeps are often portrayed as a backward kind of beekeeping because their frames cannot be removed and replaced. To illustrate the backward notion of skeps, it is often said that colonies were destroyed to remove wax and honey. I know from my own beekeeping practice, that it is straightforward to storify skeps by piling hives one upon each other to preserve movement and communication between the different stories. Surely similar methods were used by medieval beekeepers, who removed wax and honey without destroying the colony.

Over simplified accounts omit to mention that an integrated approach to beekeeping must have been essential to maintain and increase stocks year by year. A sustainable yield of wax and honey would have been essential to satisfy the insatiable demand for wax products, used for writing boards, candles, ointments, cosmetics, rituals, arts and crafts, cosmetics, polishes and so on. How else could this demand have been met without a sustainable model? These medieval beekeepers must have worked very hard to build their apiaries, hives and equipment from local biomass. It seems likely that apiary management would have been more of a full-time job and I wondered if these beekeepers paid any attention to the debris produced by honey bee colonies.

1.5 Teaching beekeeping in high schools

A group of young people asked the questions that became the inspiration for this project. Teaching beekeeping has been one of the highlights of my school teaching career that ended with retirement in 2023. I loved the way that honey bees became an engaging topic for science education. Anyone who works alongside youngsters will say that they ask the 'best' questions about honey bees, many of which seem worthy of their own PhD theses. Each year, we collected debris from the inspection screen to study using microscopes (see fig 1.7). This started as an exercise in counting *Varroa destructor* (varroa) and deciding what measures were needed to control the varroa population. During the process, we would all become enthralled by the marvellous world of 'waste'. "Mr Baxter (AKA Mr B), what's this? Why does this happen? I wonder...?" I soon realised that I was way out of my depth, and the study of bee 'waste' was put on the list of things to study after retiring from the classroom.

Debris has an important story to tell, if only we can understand its language. In the same way as human waste is a useful tool for better understanding health, bee debris has the potential to be important in understanding bee health. It does not feel like a stretch to say that bee debris has huge potential as a biological indicator of environmental change. Because the bee has great mobility and a wide flying range, contaminants will surely be brought back to the colony and bees are in effect sampling fine particulate materials from the wider environment.

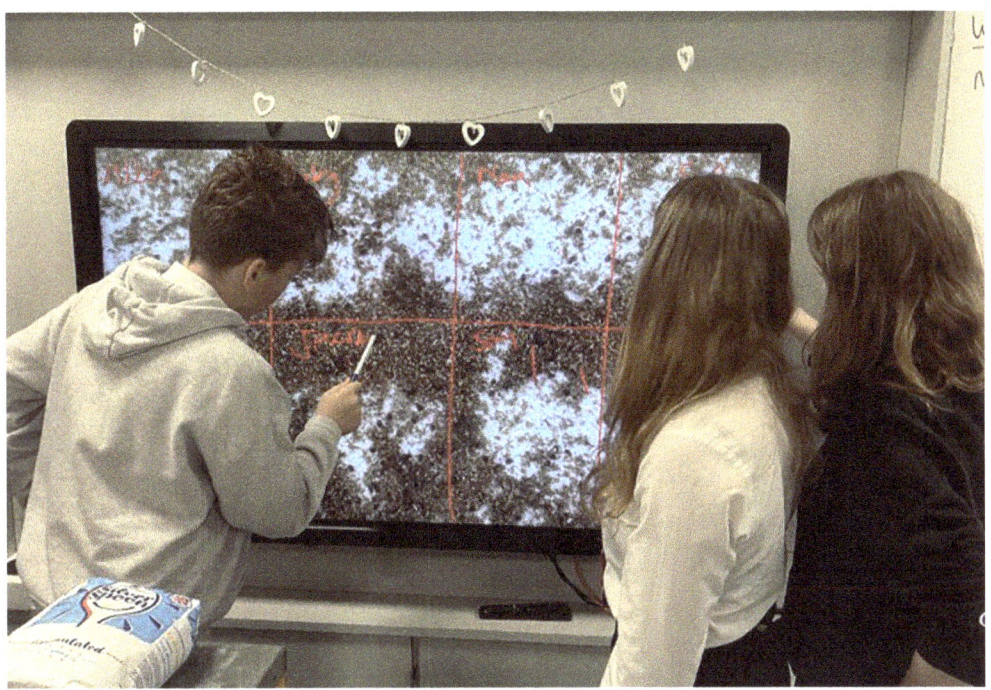
Figure 1.7 Students studying bee debris in 2020

I started making plans to create a controlled environment for a DIY research project to better understand bee debris. To be honest, I was a little daunted by it all, but, with a little bit of work every day, I built a bee lab. To date this consists of a grand shed containing three identical AZ hives, and a plan to create a controlled space for DIY research (see fig 1.8). The AZ hive is named after Anton Žnideršič (1874-1947) a Slovenian beekeeper who had seen the modern, Langstroth hives and admired the practicality of removable frames. However, like many locals, he mourned the loss of the traditional bee-skep house, a trademark of Slovenian rural culture. Žnideršič set about incorporating and improving both concepts which resulted in a hive that is now the centre piece of the Slovenian apiary. Commonly called AZ hives, they resemble kitchen cupboards, with removable frames that are accessed from the rear of the hive.

For my study it was important that the hives had inspection boards that were fully enclosed from the outside environment to protect them from both weather and rodents. I needed a set-up where the only access to the inspection board was from above. An arrangement that gave confidence that all the debris falling on the inspection screen was from the colony. The AZ hives have been designed this way with a drawer-like arrangement to inspect debris.

Figure 1.8 My home bee lab

I also thought this was the ideal time to start planning for beekeeping in later life. I have known a few beekeepers give up because of back problems due to the difficulties lifting heavy boxes. I wanted to change my hive set-up and move away from lifting heavy boxes of frames to an approach that involved managing frames rather than boxes. The AZ hives seemed like a good solution for many reasons. They also deal well with the cold weather in Slovenia, so they seemed ideal for a Scottish winter. While this Slovenian tradition was the initial inspiration, I was fascinated to later learn that there was also once a British bee-skep house tradition as illustrated by Edward Bevan's bee house (see figure 1.7). My bee house was built by the summer of 2023 and preliminary research began.

The bee house is in the middle of our family flower farm, a pesticide and herbicide-free four-acre smallholding in the Scottish Borders. It is my wife who runs the business and she is amazing at growing flowers. I help with things behind the scenes, from fixing machinery, fencing, building polytunnels, irrigation systems, workshops, septic tanks... As I write this, I am planning to move twenty tons of fresh manure that has been dropped near the front gate. There is more to running a flower farm than picking flowers in a pretty dress on a hot day, which is good news because I'd look ridiculous in a dress.

We are not traditional farmers, landowners or florists. We both grew up in cities and have needed to learn all the skills and buy all the stuff needed to create a flower farm business from scratch. It's been hard work, but a wonderful journey and we've met some amazing people along the way. The whole process happens on our small farm. The flowers are grown here; we build the soil, sow the seeds, nurture the plants, cut the flowers, condition, and arrange them. We grow what does well here, mostly outside and always without herbicides or pesticides. In a typical year we will supply more than 250 species of plants and many thousands of stems. We are proud and passionate about making a living from the business of growing flowers on our small piece of land and also turning an arable desert into an area of diverse forage for insects of all kinds. Beekeeping has been an integral part of that journey.

In time, my hope is to set up an observation hive to watch the behaviours that produce debris. In the meantime, I am reliant on personal study and correspondence with other beekeepers to better understand bee behaviour and using this to deduce what seem like plausible relationships between waste and colony behaviour.

1.6   Getting ready for home research

In the months leading up to this project I searched out and devoured literature about honey bee behaviours that can produce debris. The bibliography provides a list of most of those references.

The work of (Siefert, et al., 2021) has also helped to better understand the debris found. While studying the effect of neonicotinoids on bee behaviour, Siefert identified the need to develop ways to better understand behaviours without disturbing the colony. Part of his work resulted in the publication of 'Honey Bee Behaviours Within the Hive, Insights From Long Term Video Analysis.' This remarkable study recorded 420 brood cells for three continuous weeks. Watching these videos has been very insightful and provides valuable information about colony activity and waste. The author provides a comprehensive resource about honey bee behaviour within comb cells, thereby providing a new mode of observation for the scientific community and the general public. Thankfully, the three weeks of continual recording have been summarised into eighteen short video clips making this essential viewing for the beekeeper. In these videos it is possible to see debris being created and falling between frames. Each video lasts around one minute, creating an easily accessible resource that goes way beyond two dimensional illustrations in textbooks. All of these videos are freely available and not locked behind a paywall, which is too common and frustrating for DIY researchers like me https://journals.plos.org/plosone/article?id=10.1371/journal.pone.0247323

Having watched the videos many times, I could see value in creating a reference table that links bee behaviours illustrated by this video evidence with possible/likely waste sources. These video observations are used throughout this waste study to explain parts of the waste.

Observable behaviour recorded by video	Possible/likely waste sources
After egg laying, the eggs remain motionless until the larvae hatch. Workers successively move as deeply as possible into the cells; eggs may be pushed down towards the bottom of each cell.	Leverage is required by claws/legs working on the comb surface to move deeply into cells, resulting in wax fragments. Also, it is possible that nurse bees may reject eggs.
The heating of cells by clustering or direct incubation with the cell.	Debris associated with general bee traffic on the comb
Cell inspection involving rapid head and antennae movement	Leverage is required by claws/legs working on the comb surface for head and antennae movement.
Mandibles manipulating transparent wax scales and their transportation	Wax scales may be lost
Mandibles manipulating and transporting recycled wax	Recycled wax fragments may be lost
The use of the mandibles, legs and antennae to manipulate wax to form brood and honey caps	Wax fragments can be lost. Also, leverage is required by claws/legs working on the comb surface to manipulate the cappings
Crawling into cells to store honey	Leverage is required by claws/legs working on the comb surface to store honey.
During pollen storage, the rear legs enter the cell first. The worker holds onto the upper cell wall and uses its legs to brush off the pollen into the cell. Leg cleaning and pollen-pushing is repeated several times until the legs are cleaned	Vigorous Leverage is required by claws/legs working on the comb surface. Also pollen and hairs may be lost in the grooming and pollen packing process.
The mechanical cleansing of surfaces within a hive is also known as the 'rocking movement'. The workers' mandibles and front legs are used as scrapers.	Debris lost during the scraping process
Removal of fungi and parasites that endanger the colony's survival.	Fungi and parasites discarded
Grooming is a very vigorous activity. The mandibles and legs are used to clean all bees	Discarding of foreign material collected while foraging
Cannibalism of larvae, pupae and parasites	Discarded body parts

Preparatory work started in the summer of 2023. I looked at lots of bee debris and asked, what shall I observe and measure? I was limited to simple observations using a mobile phone camera, microscope, a centrifuge and a mass balance that I purchased for this project. The starting list included recording month by month, data and images that show:

- The total weight of debris and by quadrat
- The number of body hairs in the debris
- The number of chalkbrood structures in the debris
- The pollen species found in the debris
- Repeat photo-point monitoring of the debris

Chapter 2 Methods used

2.1 Photopoint monitoring using a mobile phone camera

Photopoint monitoring consists of repeat photography of the inspection board over a pre-determined time period; it is an easy, yet very effective, method of monitoring debris and colony activity. I used an old iPhone camera, but any modern mobile device will probably take better images than those included in this book

Through trial and error, I have learned that it is best to take photos early in the morning, late in the afternoon, or on days when the sun is less intense. This reduces shadows and harsh glare in the photos and makes identifying parts much easier. Creating a gallery record of pictures helps to spot and track changes over time. These pictures will be time-stamped, making storing, retrieving and sequencing images much easier. The whole process only takes a few minutes, a good use of any beekeeper's time.

It's incredible what can be seen by zooming in on these images using the most basic mobile phone camera. However, things will be even clearer when viewed on a large screen. Also, there are many kinds of software that can help with the automated counting of objects in photographs. I have not tried these but have been interested in these possibilities.

How and when photographs are taken will be determined by the goals of the beekeeper. In this study photoint monitoring was carried out in two different ways.

The initial plan was to record photographs (similar to figure 2.1) for a 28-day period for each month. Mostly this worked well, but in the summer months it became difficult to keep track of changes due a large build-up of debris. For this reason, in the summer months photographs were taken every 14 days.

Opportunistic photos also provided interesting information about hive interventions or events in the colony. An example of this happened in June after catching a cast swarm. I wanted to compare the debris created by the colony before the queen had been mated and after the subsequent changes in colony development. I have also created 'opportunistic' records to examine the debris produced after hive inspections and post-feeding. There are of course many other potential opportunities for this kind of record keeping. For example, will the debris change after a treatment of agrochemicals in nearby fields, due to hive interventions, during food shortage events, due to a wide range of different bee diseases etc?

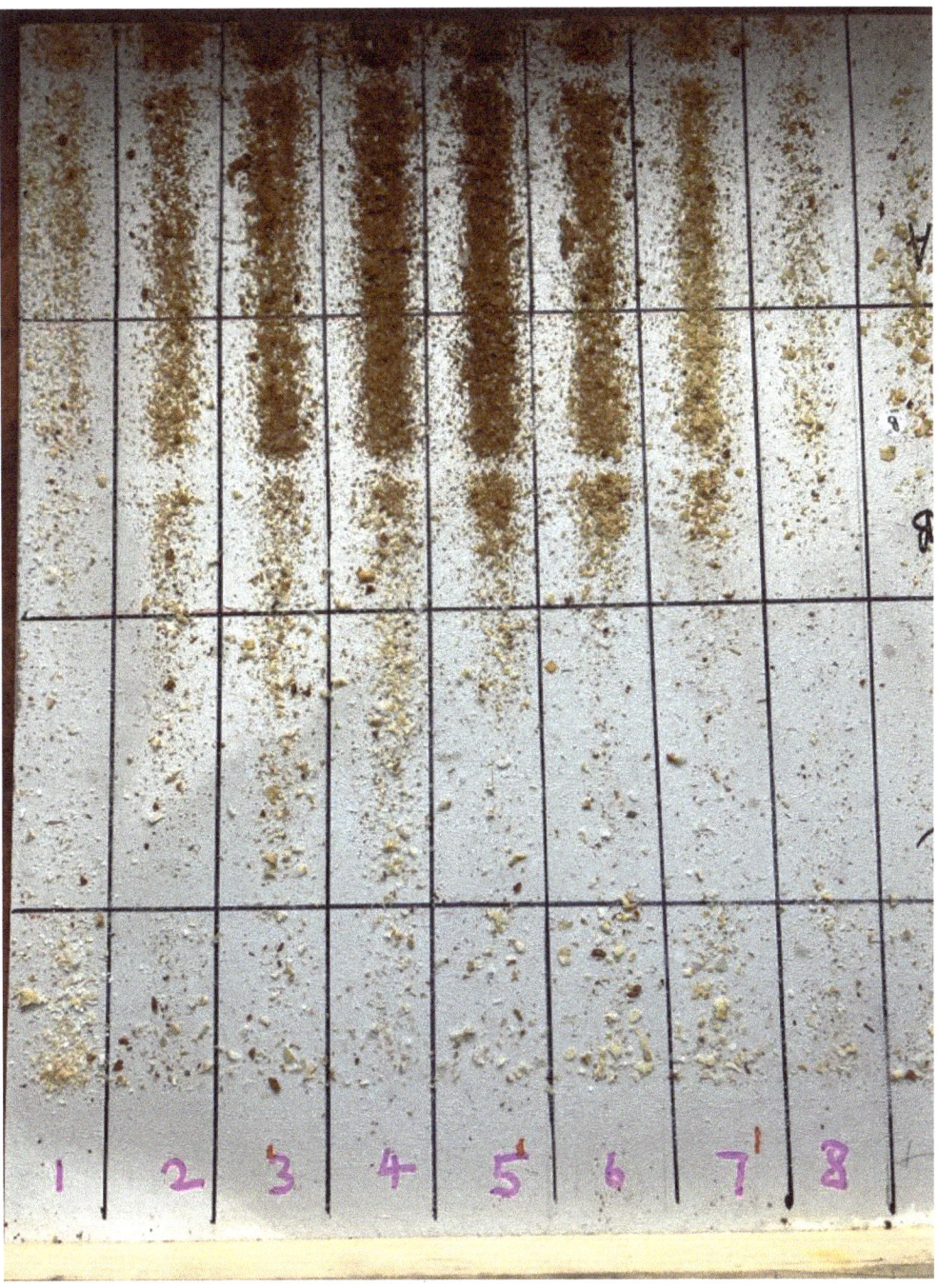

Figure 2.1 Repeat photopoint monitoring

2.2 Weighing debris and creating heat maps

Two different methods for weighing debris have been used in this study.

The first involved dividing the debris board into 36 equal quadrats and weighing the total amount of debris from each quadrat (see figure 2.2). This method was used to create the heat maps shown in chapter 3.

The second method involved weighing the debris on the inspection board for each month.

Figure 2.2 Weighing by quadrat

The digital scales used for this project had a claimed accuracy value of 0.001g. Thankfully, the detection of such small differences in weight wasn't required because these scales struggled to give consistent accurate readings at this level of precision. For this study, measurement to within 0.01g (one hundredth of a gram) was considered more than adequate and the scales were suitable for the task. Similar scales can be bought relatively cheaply.

To create the heat maps shown in figure 3.2 the debris weight from each quadrat was added to a table in Excel. The software has a useful feature that can create a heat map to represent the relative amounts of debris in each quadrat. As the intensity of the red colour (heat) increases so does the weight of the debris in that area, which in turn suggests an increase in colony activity. The assumption is that more debris equals more activity, and that debris weight is a proxy indicator of activity.

This Excel wizard creates heat maps with four shades of red. These shades represent below 25% from the mean weight, between 25 and 50%, above 50 and 75% and above 75% from the mean weight in each quadrant sampled. Creating these heat maps has been an interesting experiment to visualise how the debris changes by month. Although at present I'm not sure what extra benefit they provide above simple monthly photographs of the inspection board as shown each month.

2.3 Counting using microscopy - estimating the number of parts from the debris

A microscope was used to count and estimate the number of hairs, chalkbrood, and fibres. This method of microscopy includes three main stages. The first is sample preparation, second is choosing an appropriate magnification and lastly using a standardised systematic method to count as accurately as possible. The detail of each of these stages may vary according to the nature of the study. Figures 2.3 to figure 2.9 describe the steps used in this study. In preliminary work, it was found that a solvent (white spirit) helped to reveal and count its constituent parts for counting purposes. The graphs shown in figures 3.1 to 3.4 have been produced using this counting method.

The simple rule is that all the 'counted' numbers described in this book are either the average mean value of 6 slide samples (shown in figures 2.3 to 2.9) or are an estimation based upon that value. The standard deviation from each mean was also calculated. These are shown by the error bars displayed by each graph in section 3. These deviation values were used to make judgements about data reliability and identify methodological improvements. This process was ongoing throughout the study.

Separate larger parts

Use a suitable tool (fine paintbrush, tweezers, needle) to remove larger items from the debris. The picture shows pollen lumps, mummified chalkbrood, wax capping's and propolis. The composition of these larger lumps will change from month to month and hive to hive. After removing the larger parts mix into a consistent looking mixture.

Label and store these samples for microscopy

Figure 2.3 Sample preparation stage 1

Samples for microscopy

Measure out three samples from the mixture.

Sample 1 = 0.05g
Sample 2 = 0.05g
Sample 3 = 0.05g

Figure 2.4 Sample preparation stage 2

Six slides from each sample

Use a fine paintbrush to divide each 0.05g debris sample into six equal lumps.

(Each of the six sub samples will weigh approximately 0.008g).

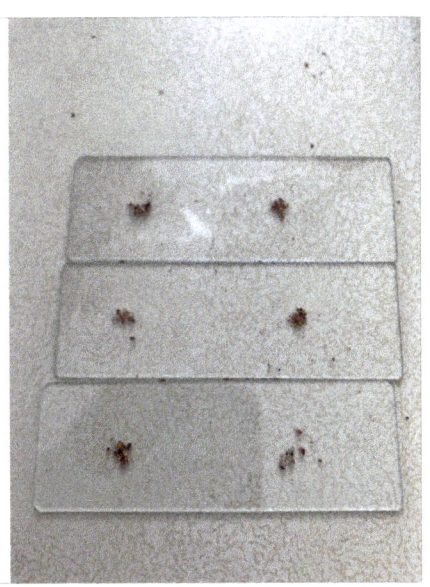

Figure 2.5 Sample preparation stage 3

Dissolve the wax

Use a fine dropper to add a suitable solvent to each sample.

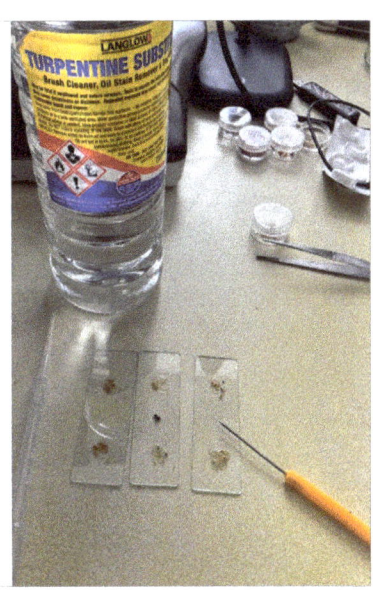

Figure 2.6 Sample preparation stage 4

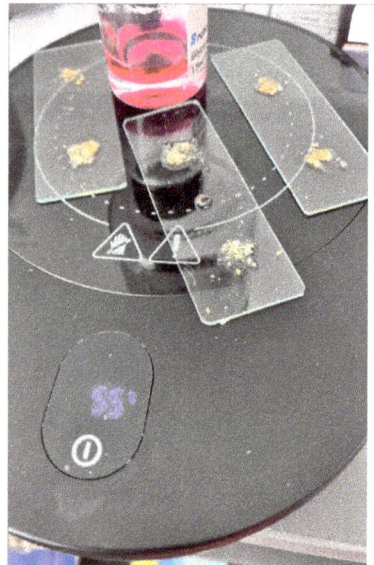

Evaporate the solvent

Place the slides on a hotplate set at 50 degrees Celsius to evaporate off the white spirit. Use the same hot plate to melt glycerine jelly.

Figure 2.7 Evaporate the solvent

Stain each sample

Add a drop of stained glycerine jelly onto each of the slide samples. Carefully lower the cover slip on top of each sample. Use a lint free fabric to clean any excess and avoid **any** contamination with the microscope.

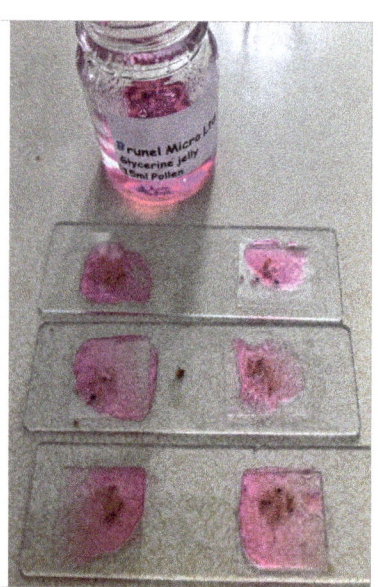

Figure 2.8 Stain the sample

Microscopy

Start by using the 10x objective. In this study the number of hairs, microfibres and chalkbrood cysts were counted on each slide. A mean was then calculated.

The mean value was then multiplied by the total number of possible samples in each month's debris. This mean value provided an estimation of the total number for each variable. A standard deviation was then calculated to better understand areas of error and to look for improvements.

Figure 2.9 Microscopy

2.10 Using microscopy to identify pollen

If you want to identify pollen you will need to become proficient at describing the double-layered cell wall that forms the external structure of each pollen grain. You will also need to learn how to measure the diameter of each pollen grain using a graticule. The internal cell wall is called an intine, while the external one is called an exine. The visible features of these exine structures are used to help identify the plant species along with the shape, texture, colour and size of each pollen grain. It is a simple process, but making a confident positive identification takes a considerable amount of research, training, skill, practice and a good mentor. One challenge is caused by using a microscope, which creates a very narrow depth of field. This means that only a very thin layer of the pollen is in focus at any one time, which can make it difficult to observe all the 3D pollen features required for a successful identification. To overcome this several views of each pollen grain are often required. Techniques like focus stacking, where images of different focal depths are combined to create a single image are useful. These techniques have been used to identify the pollens described in this book.

I am very grateful that Christine Coulsting, a co-author of Margaret Adams' book Pollen Grains & Honeydew was on hand to help me improve my identification skills.

Recommended references include

- Pollen Identification for Beekeepers (Sawyer, 1981) is essential reading for understanding the features of pollen. He defines much of the vocabulary to explain pollen features, as well as establishing the acknowledged Pollen Identification Key.
- Pollen Microscopy (Chapman, 2018) shares techniques for microscopy accompanied with wonderful illustrations and drawings of pollen structures.
- The Palynological Online Database is essential. It provides SEM photographs of pollen grain features and from light microscopy.
- Pollen Grains and Honeydew (Adams, 2021) describes in detail how to prepare samples, slides and carry out microscopy to identify pollen structures

Chapter 3 Graphs and data visualisation

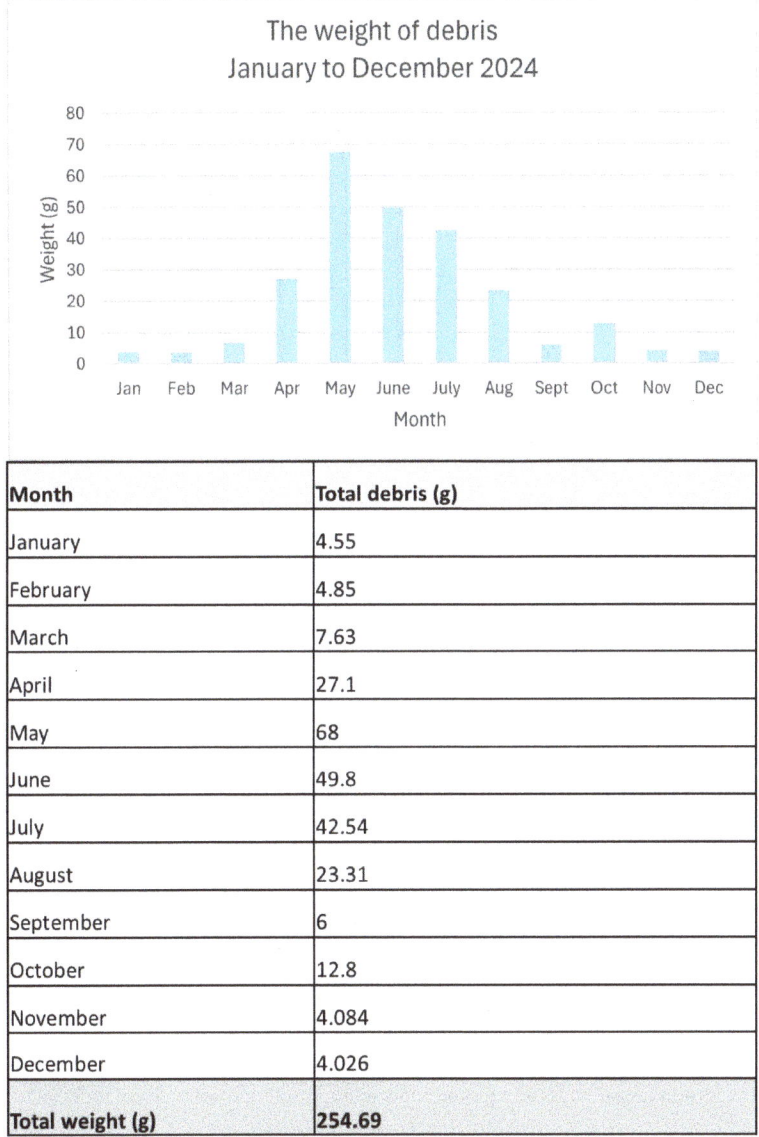

Month	Total debris (g)
January	4.55
February	4.85
March	7.63
April	27.1
May	68
June	49.8
July	42.54
August	23.31
September	6
October	12.8
November	4.084
December	4.026
Total weight (g)	**254.69**

Figure 3.1 Debris weight

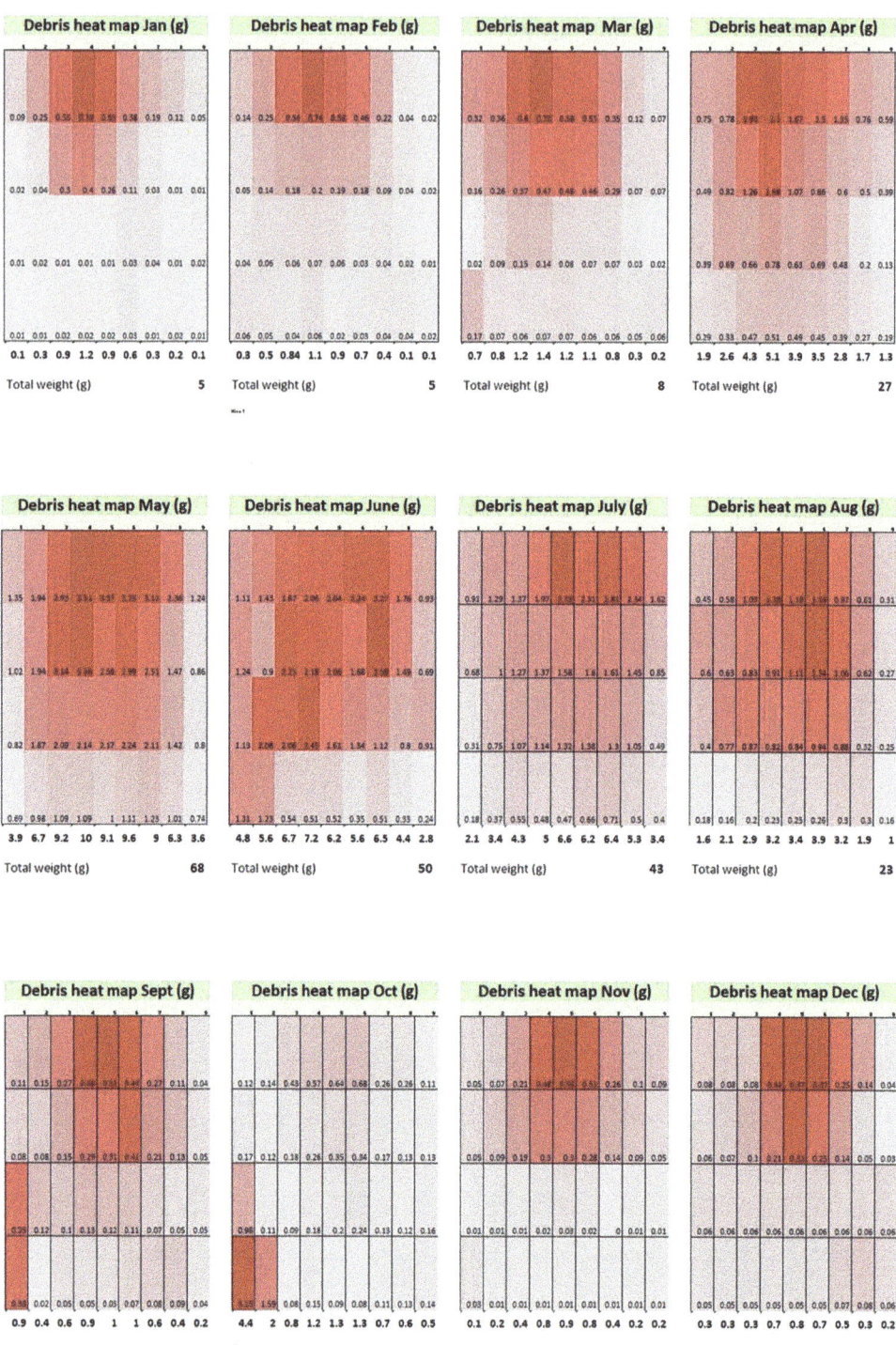

Figure 3.2 Debris heat maps

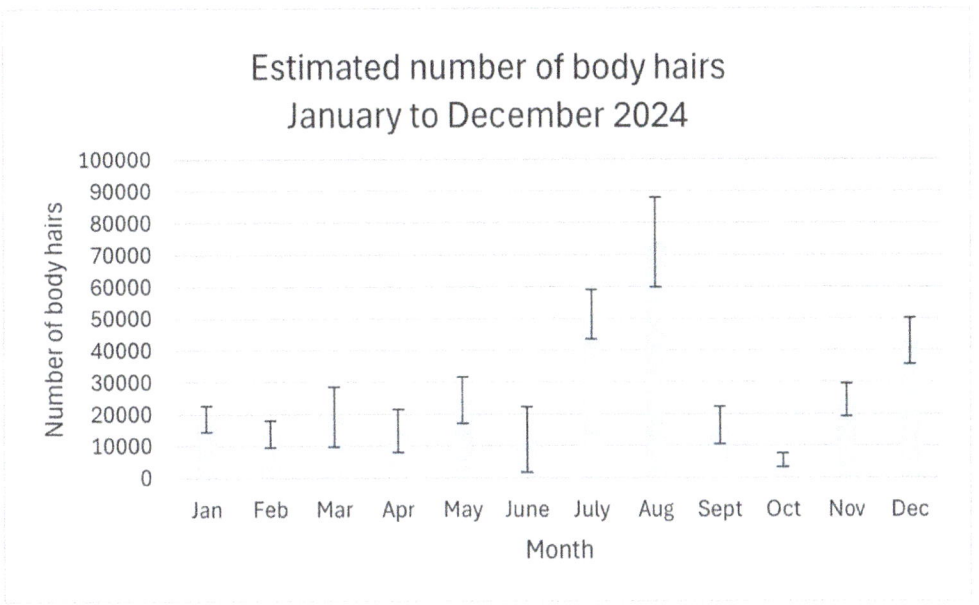

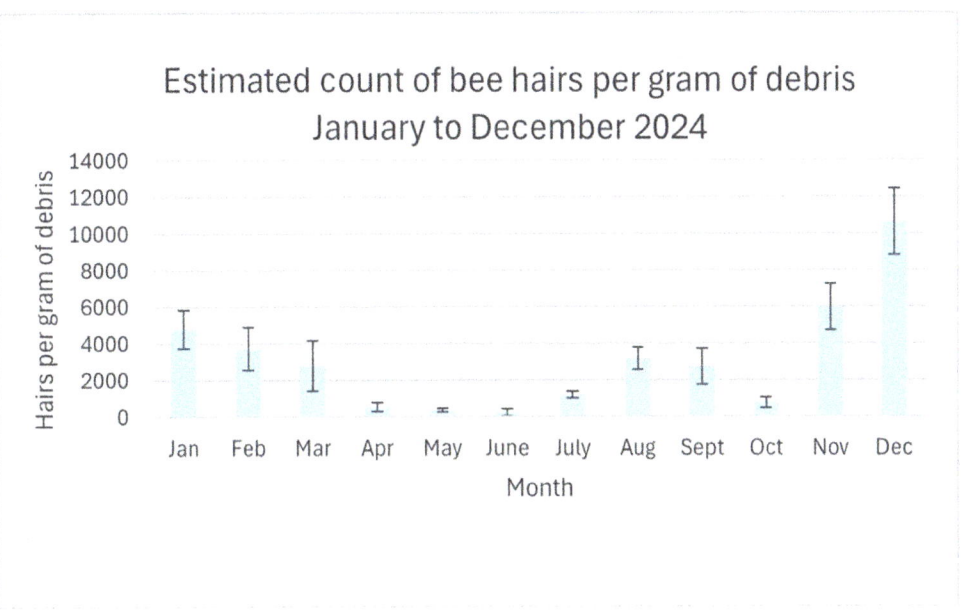

Figure 3.3 Bee hairs

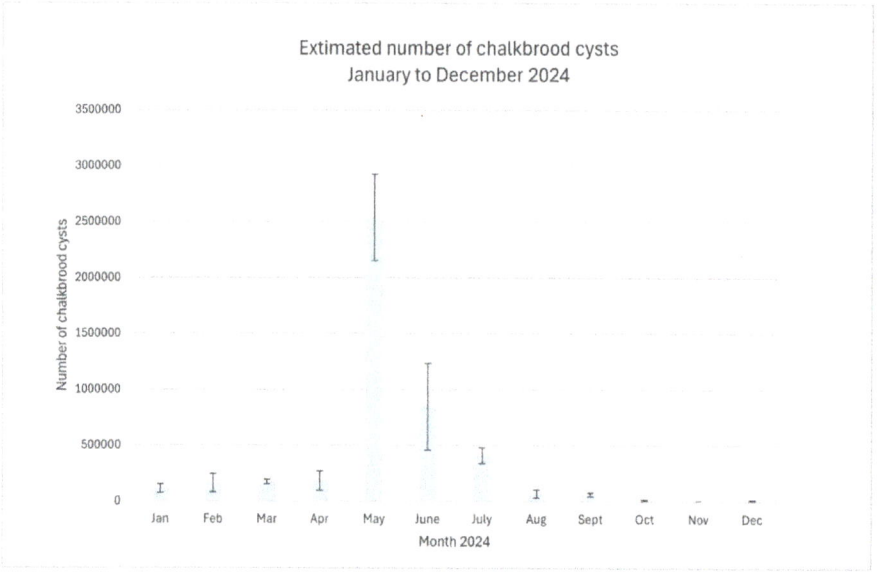

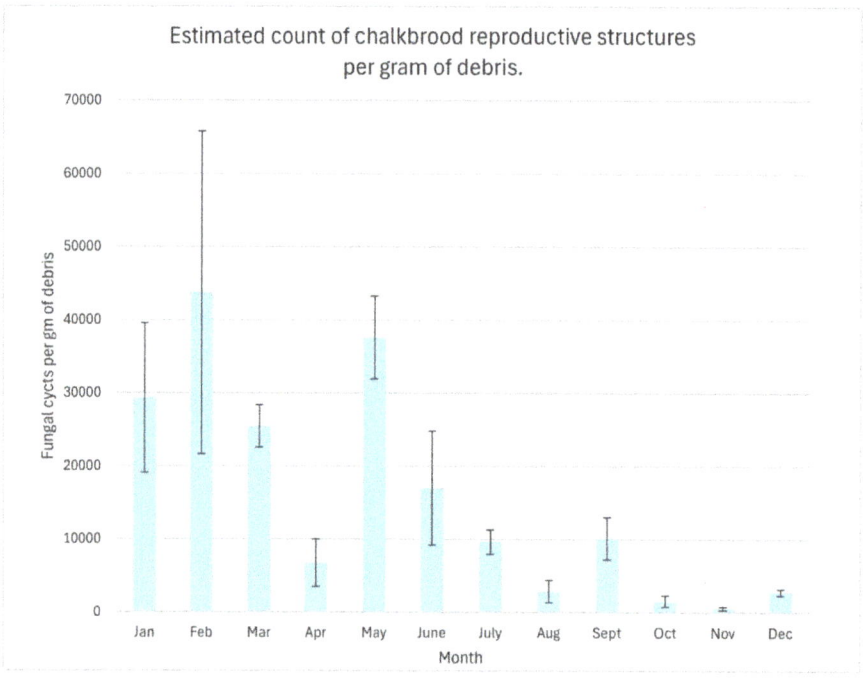

Figure 3.4 Chalkbrood cysts

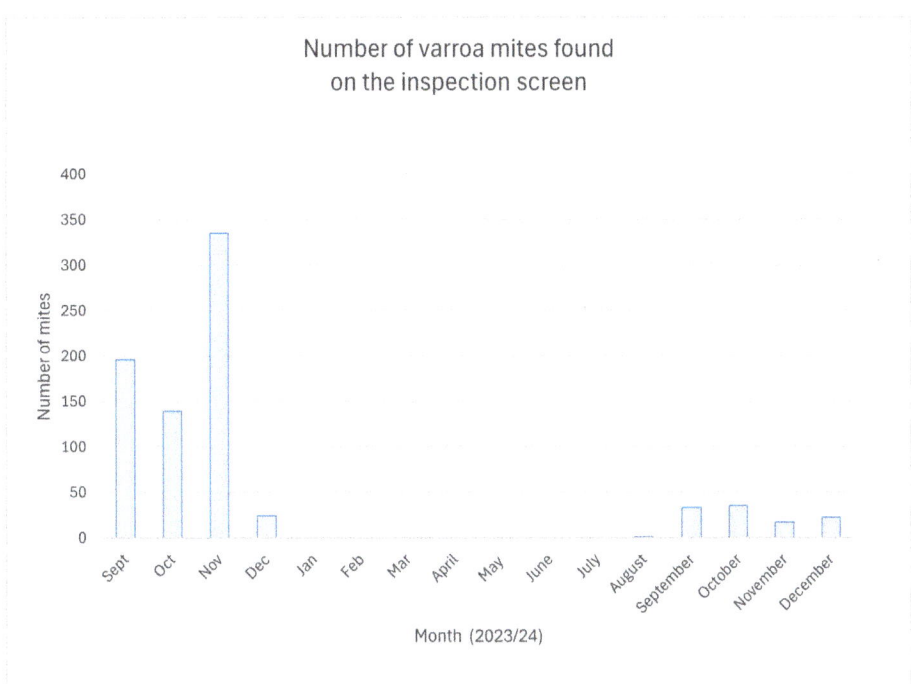

Figure 3.5 Varroa mites

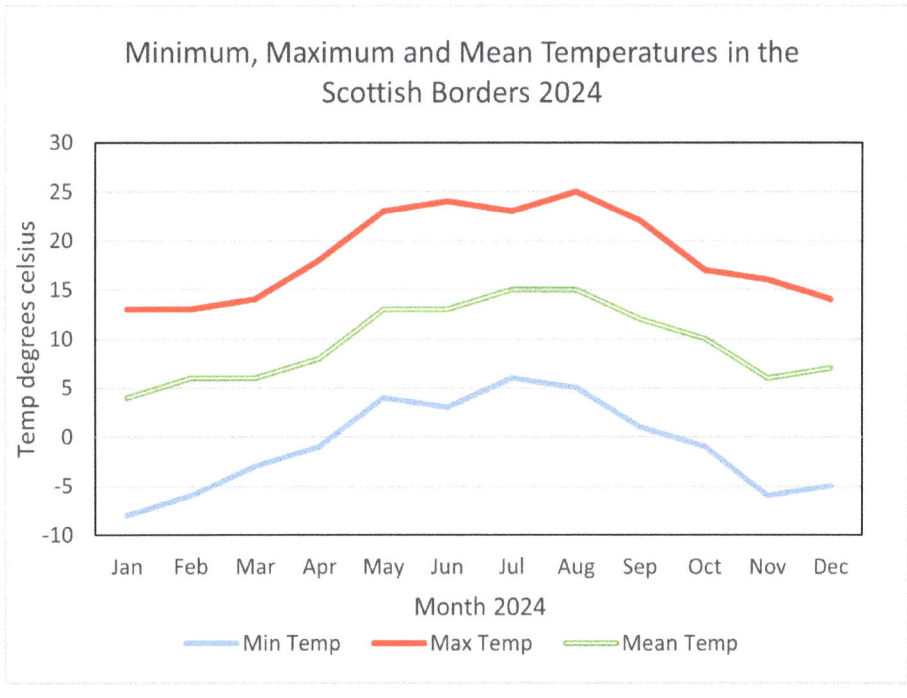

Figure 3.6 Minimum, maximum and mean temperatures in The Scottish Borders.

Chapter 4   Twelve months of bee debris from Mill Pond Flower Farm apiary

Before writing a diary the first task is to agree on some kind of structure. As a starter each month will include:

- A picture of the inspection board showing 28 days of debris
- A picture showing the pollen species found in each month's debris
- Pictures of other finds of interest
- A discussion about emerging trends, patterns and finds as they happen.

4.1 January

Nearly a gold dance

Honey bee debris collected between 1st and 28th January

Minimum temperature -8°C

Maximum temperature 13°C

Mean temperature 4°C

On 1st January, I slid a clean inspection screen underneath the brood chamber. I have been telling my bees about this project for the last 6 months, but I doubt that they will remember that on the 28th day of each month that the inspection screen will be removed and the debris collected for analysis.

Figure 4.1.1 shows the pattern of debris for January. The debris is found in the southernmost aspect of the hive and nearest to the hive entrance. This 4.55g of debris is just over 1g of debris falling from the colony each week. It seems reasonable to predict that January will contain less debris than any other month as it is likely that mid-winter colony activity will be at lowest value - an optimum size to survive the shortest days and coldest temperatures of a Scottish winter. Because this is January and the start of this project it is difficult to comment further on this single value, but it will be fascinating to compare this value with future values and the trends as they

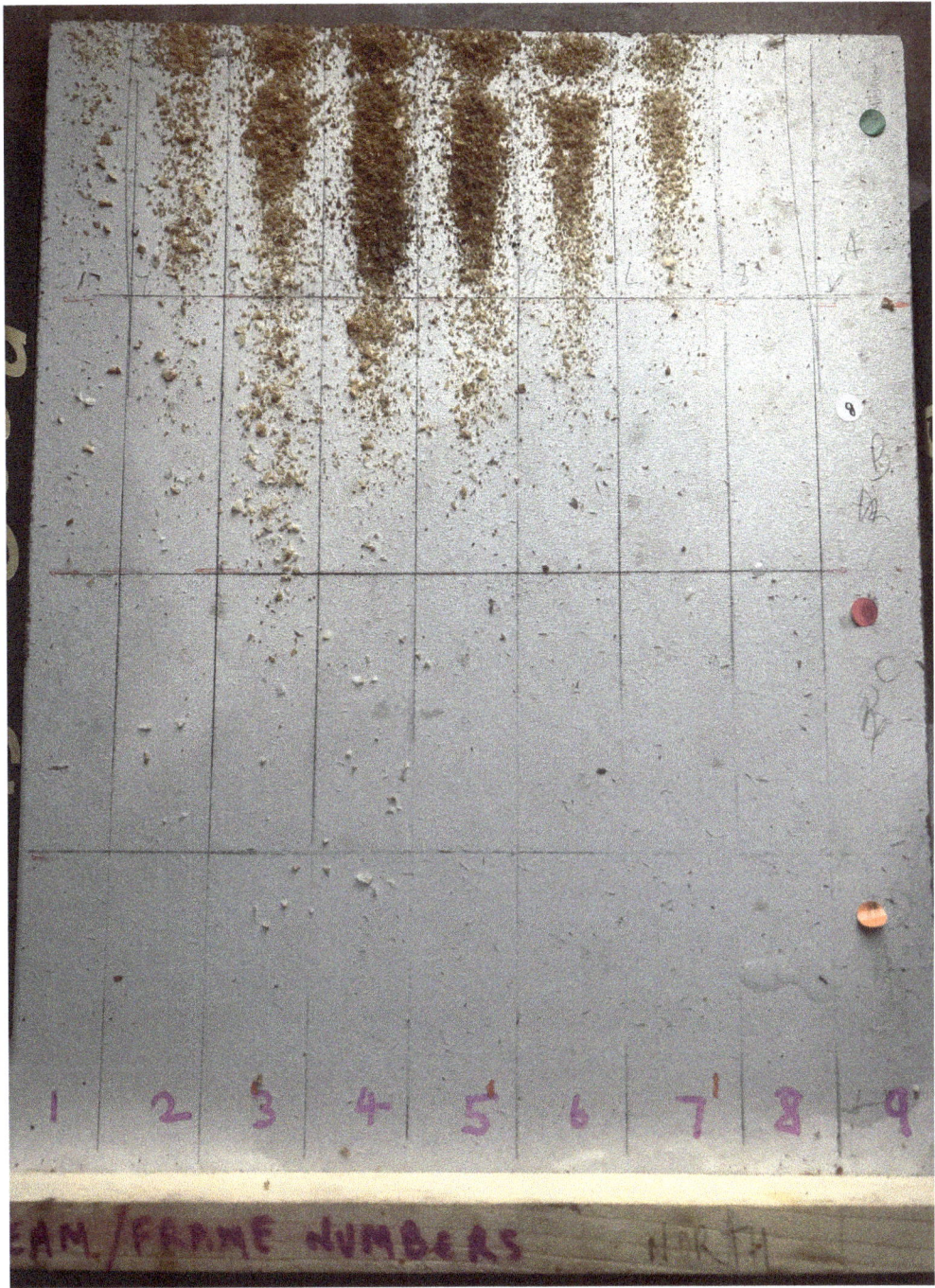

Figure 4.1.1 January debris

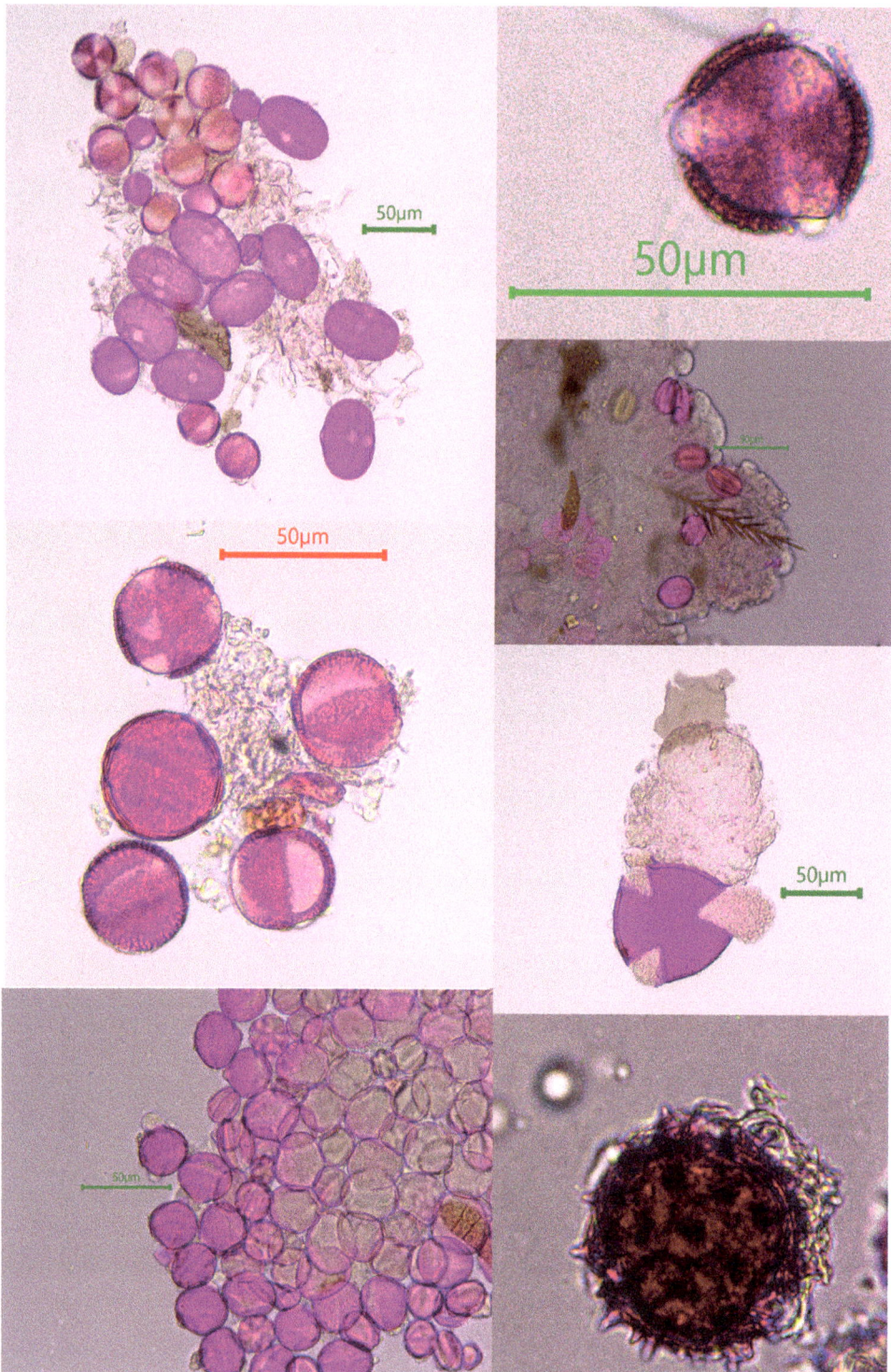

Figure 4.1.2 Pollen found in January debris

Pollen in the debris

Six different flower species were found in the January debris. All of these were last in bloom in the summer of 2023, almost 6 months earlier.

These pollen grains included field beans, radish, rapeseed, phacelia, rosebay willow herb and a member of the Asteraceae family. This is the largest family of flowering plants and they make up a large proportion of the flowers grown on our flower farm. This family is commonly known as the aster, daisy, composite, or sunflower family. Many of the pollen grains of the Asteraceae family share similar 'spiky' characteristics which can make identification problematic. I believe the spiky Asteraceae pollen shown in figure 4.1.2 is from a thistle flower. The half an acre of thistles growing outside the flower farm gate make it highly likely that my bees will forage on these.

Left   Top = field beans and rapeseed, second down = rapeseed, bottom = phacelia

Right Top = rapeseed, second down = phacelia, third down = rosebay willowherb, bottom = Asteraceae

With limited forage availability in January and it being too cold for flights, it is not surprising to find evidence from last year's pollen stores in the waste. It is a useful reminder of the importance of a balanced diet for bee health.

The vital role of honey bee colony nutrition is well illustrated by Kahn et al. (2022). This study looked at the effect of nutrition on queen fertility and colony performance. The research team fed honey bee colonies different diets and measured the rate of egg laying and the area of brood sealed. The results showed that colonies fed a mixture of pollen and sugar solution produced queens that laid significantly more eggs and produced a greater area of sealed brood, compared with colonies that were fed only sugar solution.

Anyone who keeps animals must ensure that nutritional needs are being met. We often use ingredient labels to assess the nutritional value before buying food for ourselves or animals. Floral forage however is very variable across localities and time. All apiaries will have a unique floral diet, depending on the forage available. It is the responsibility of beekeepers to do their homework to understand the quality of the forage available.

I remember as a new beekeeper quickly developing a laser-like focus looking for nectar and pollen producing plants near my apiary. Also, trying to identify the pollen at the hive entrance and fitting a pollen trap on the hive to collect pollen for microscopy. Developing a keen sense of environmental awareness really is one of the highlights of becoming a beekeeper.

After completing an introduction to beekeeping course, I soon understood that while I may live in a lovely remote rural spot in the Scottish Borders, the forage availability was poor for bees and other pollinators. The 1000s of acres of mainly potatoes, wheat and barley, while putting food on the table for people, did very little for pollinating insects. With the odd field of yellow rapeseed, it became a cycle of boom and bust for my honey bees and they found it hard to survive without human intervention and a reliance on supplemental feeding.

Twelve years later, having established a flower farm on our smallholding and the start of a new farming philosophy emerging in the fields surrounding my apiary, things are starting to improve and home is becoming a much healthier place for honey bees and other pollinators. I really wish that I had started this debris study twelve years ago to have a data record for comparison, but it is impossible to imagine that the same pollen species would have been found in the waste. Each year, I need to feed my bees less, and each year I look forward to planting more flowering plants to produce a wider range of seasonal forage.

I am learning that the study of debris provides a very useful way to understand what honey bees are actually eating month by month, in effect creating a time stamped ingredient list of a unique floral recipe. It will be interesting to watch how these change in future months and the years to come.

Nearly a gold dance - a golden egg

The holy grail of metal detecting is gold and many have a choreographed dance move ready for when gold strikes. I am still waiting and am rubbish at dancing, so cannot imagine that happening, although people say that you never know what will happen when gold fever strikes. A few finds have fooled me, one was a shiny gold top from a milk bottle, common in the 1960s for full fat milk, for a moment I thought it was a gold coin.

I nearly started to dance, when I found this egg on the waste board (see figure 4.1.3). Beekeepers are trained to recognise single eggs at the bottom of cells, a gold star indicator that there is a laying queen. I was delighted and posted a picture to my Facebook page saying something like, the queen has lost her egg and how happy I was knowing that the colony was queen right. It was not long before people commented and made different conclusions about what may be happening. This is just like metal detecting I thought.

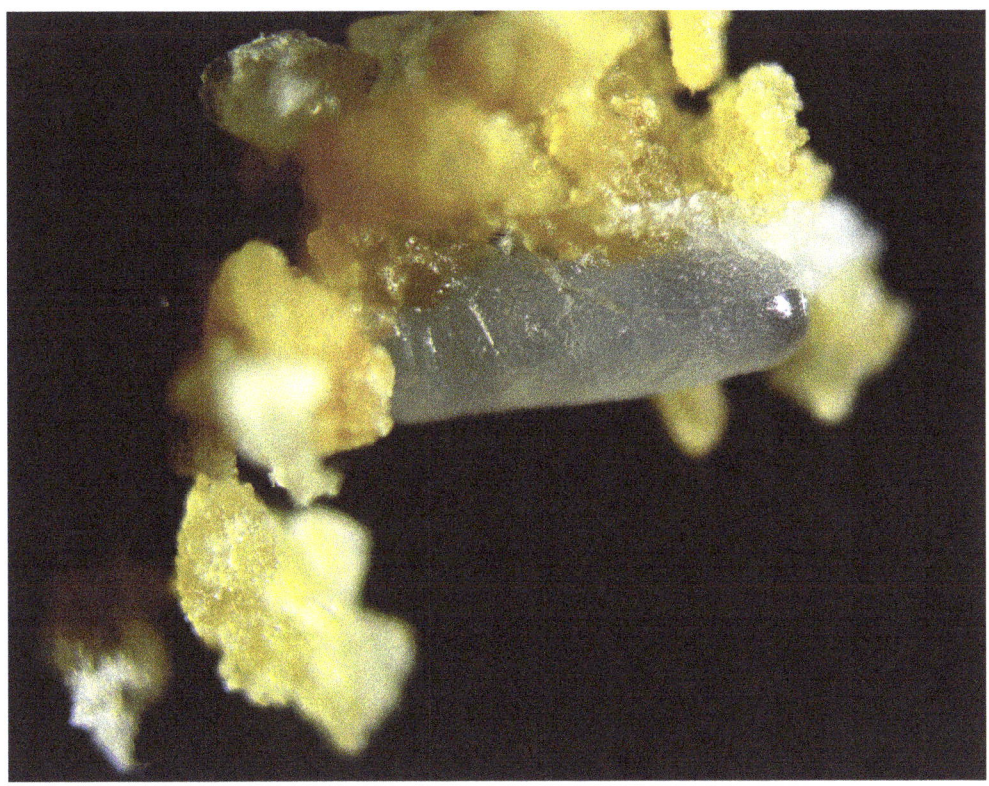

Figure 4.1.3 A golden egg

On Saturdays, members of my local club of detectorists post their finds and we discuss and help each other to identify and contextualise each other's treasure. You've got to be humble and be happy to be wrong, there is always someone who knows more than you. It's amazing how many times that I have posted a tiny fragment of an old hammered coin, only for someone to identify when and where it was made. It's the same in all scientific endeavour, an unshakable attribute that should underlie all research endeavours is humility and the quality of embracing error.

I have learned that I was probably wrong by thinking that the egg had been laid by a queen. I did not know that worker bees lay eggs in queen-right colonies (albeit in very small numbers). (Visscher, 1989) estimates that 0.12% of drones are produced from eggs laid by workers in queen right colonies. Also, (Pirk, et al., 2004) describe the strategies used by the colony to deal with workers in a queen right colony. In some cases, egg laying workers will be attacked by other workers and other workers effectively police the eggs, by removing eggs that have been laid by workers.

The next day I found several more eggs, some in pairs on the waste board (see figure 4.1.4). I may not have struck gold but have had fun learning and know a little bit more than I did yesterday. There may of course be other explanations for the presence of these eggs. Also, it is a mystery why these eggs have been discarded, rather than eaten.

Figure 4.1.4 A pair of eggs

Human made fibres

Ann Chilcott, one of the editors of this book, asked if I was finding any microplastics. I knew that I was finding something that looked like fibres, but didn't know how to identify them. I had dismissed the idea of studying these, as there was already plenty to do. However, I kept finding them entangled in waxy lumps. Were these man-made, plant tissue or other kinds of fibre? I was doing a month-by-month microscopy study anyway, so why not add a monthly catalogue of nanofibers too.

It does not feel like a stretch of the imagination to say that waste from honey bee colonies has the potential to be a biological indicator of environmental change. Each forager has many thousands of hairs that have evolved over millions of years, partly to favour the collection of small particles of food. Because a bee has great mobility and a wide flying range, contaminants must surely be brought back to the colony. It is likely that these will be processed by the bees and evidenced in the debris on the inspection board. Debris has the potential to tell a much bigger story about environmental pollutants. If only we can learn how to read this debris.

I needed to find a way to quantify the numbers of fibres sampled each month. The same method described in section 2 used for counting hairs and chalkbrood structures seemed to work well. The headline result for January was an estimated 16800 fibres present in the debris. Figure 4.1.5 shows just a few of these fibres that were present in the studied samples.

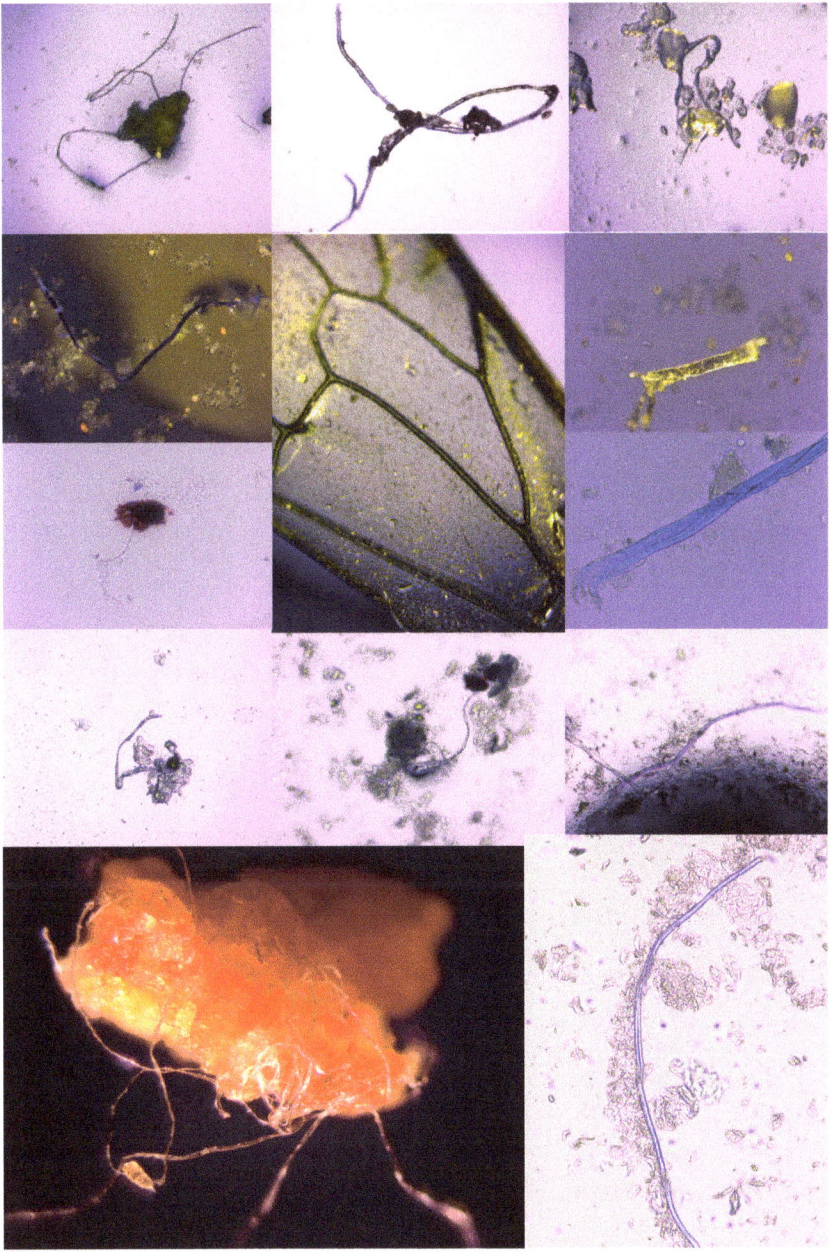

Figure 4.1.5 Fibres found in the debris

I started reading about the mechanics of bee flight and foraging. I dreamed of bee flight scenarios, with wings beating over 200 x per second, twisting and rotating to best manoeuvre. I tried to visualise the mini vortices created by the wings, acting like micro tornados, kicking up pollen and other small particles of just the right size and shape that could stick to body hairs.

It is well known that helicopters create huge clouds of particulates when landing or taking off, causing dense clouds of particulates that are thrown into the sweep of the rotors. Operating in these conditions is dangerous for the pilot and a major cause of accidents during military operations. It seems probable that foraging bees return to the colony with whatever particulates were present in the environment at the time of flight. That the bee becomes a bio-sampler of fine particulates with a favourable structure and electrical charge. Also, like the helicopter analogy a maintenance crew is required to clean and service the helicopter before its next flight. Perhaps the fibres found in the debris are a product of this grooming behaviour.

The wing picture shows fibres clearly attached to a wing hair. This reminded me of foragers covered in pollen on a hot summer day. Many of the waxy lumps found on the inspection board seem to be a product of grooming and waste processing, with each waxy lump containing many bee parts, hairs, nanofibres, pollens, and fungal structures.

The presence of so many fibres in the waste prompts so many questions, and you may also have your own?

How does the fibre profile vary between apiaries?

Is there a way to identify natural, man-made fibres and microplastics?

Which month will contain the greatest number of fibres?

Is the packaging of fibres into waste an example of hygienic behaviour during grooming?

What is the grooming burden for a colony needing to process nanofibres?

What is the source of these fibres and how do they get into a hive?

How do these fibres affect the colony?

What can these fibres tell us about the wider environment?

Many of these questions are way beyond the scope of this observational study, but they are now on the 'to study' list and for background reading, to mull over during the life of this project. Perhaps the monthly fibre estimates, and microscopy will provide more information about the presence of these fibres in bee hives.

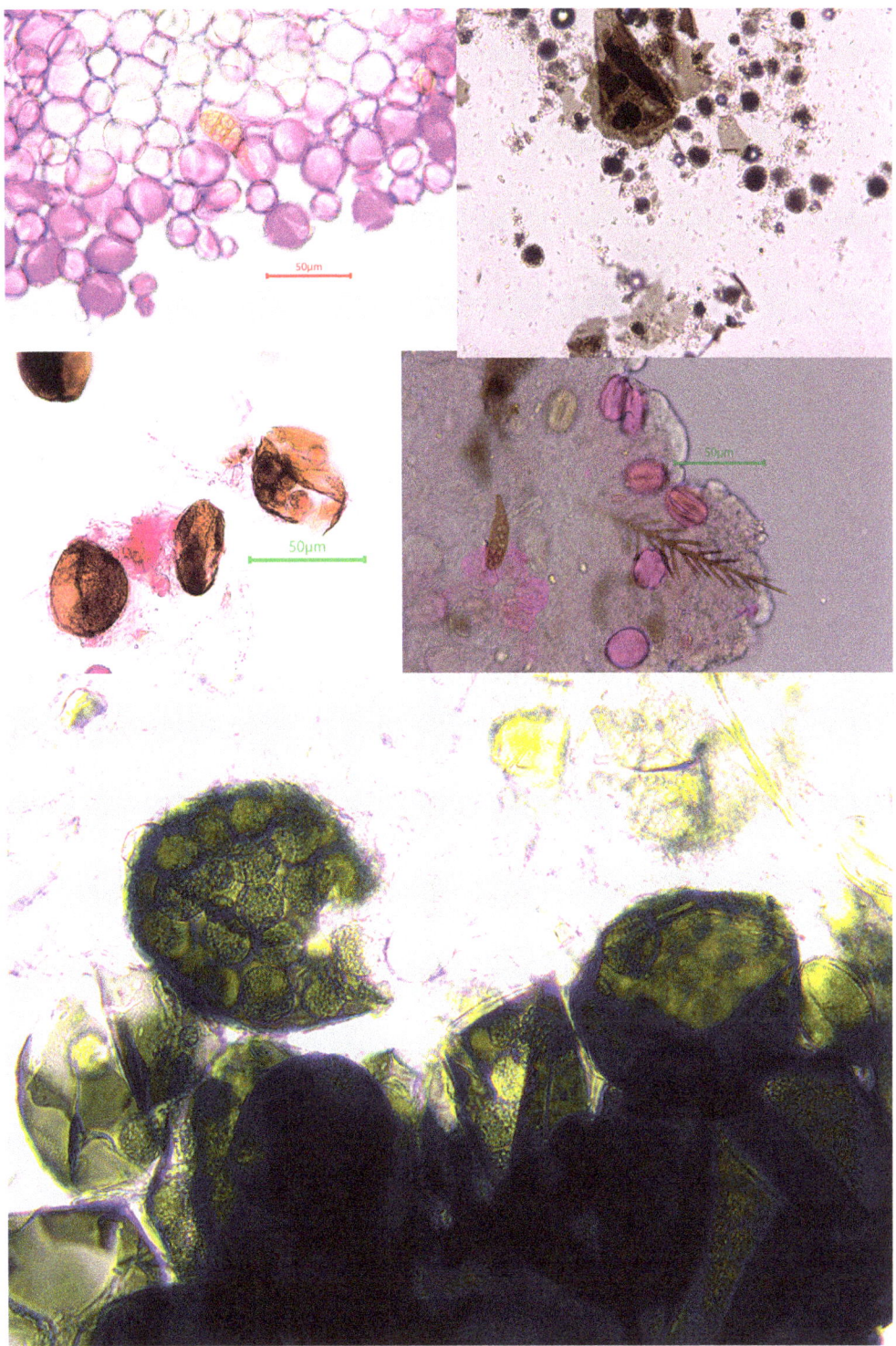

Figure 4.1.6 Chalkbrood structures and a species of Alternaria (top left amongst pollen grains)

Fungal species found in the debris

Two different fungal species were found in the debris (see figure 4.1.6). A species of Alternaria and chalkbrood (Ascosophaera apis). It's the first time I have seen any of the 299 known species of Alternaria genus. Apparently, they seem to survive almost everywhere and are responsible for 20% of food spoilage. I wonder what role these fungi play in honey bee colonies?

Chalkbrood cysts look to me like a bag full of footballs, the footballs or spore balls (around 20 µm) then in turn break open to release spores (Around 5 µm).

It is fun adding a drop of water onto dry cysts and watching. Within seconds the spore balls are released and they burst open releasing their spore loads. Millions, or is it billions, of tiny spores with all they need for reproduction are set free. Little wonder they are so effective at infecting bee colonies in humid environments. I wished that I had discovered this practical skill when I worked in schools - students would have had fun too and it would have created lots of great discussion about osmosis, beehive humidity and more.

Using the methods described in chapter 2, the headline result is that there were an estimated 11000 chalkbrood cysts present in this debris. There must be billions or is it trillions of spores right there. A reminder of why it is important to regularly clean the inspection board. It is feasible that these are a consequence of hygienic behaviour when workers eject these structures from their nests. Condensed cyst lumps appear a dark colour in the waste and are easily visible by eye. The lump shown by figure 4.1.7 will contain hundreds of these structures.

Figure 4.1.7 A lump of chalkbrood found in the debris

Body hairs.

The same method was used to estimate the numbers of body hairs in the debris. The headline result was that there were more than 18000 bee body hairs in the debris (see figure 3.3). Bee hairs are a very common find that require an explanation. The video Honey Bee Hygiene: Allo-grooming (Siefert et al.) is helpful. If a picture says 1000 words, then this one-minute video must say at least 10000. A large screen is recommended for viewing, so that the bee's mouth parts and clawed feet can be seen in action.

The grooming process starts when worker bees invite others for grooming. This 'grooming invitation dance' involves rapid leg moments and much body bending. Flo, our scruffy spaniel dog, does something similar when she wants a brush. She runs about madly and rubs her body against the furniture. Flo is a spoiled farm dog and is about outside most days. She is expert at collecting all kinds of stuff in her hair; seeds, pollens, soil, ticks, twigs, plants, mites, and so on. No doubt, the debris 'collected' by dogs would tell you a great deal about their lifestyles.

The video shows the honey bee wanting to be groomed, using its claws to grab and firmly hold onto the edges of a section of empty wax comb. The grooming bee uses its mandibles to bite and nibble, paying close attention to the connecting joint between the wing and body of the honey bee. After a period of close nibbling the bee uses considerable force to move its head back and can then be seen chewing. At the same time all six legs can be seen independently grabbing, brushing and pulling at clumps of body hair all over the body. Honey bee grooming is a vigorous activity, the video helps to understand the relative strength of honey bee legs and the neck muscles to pull with such force. Watching the video makes it much easier to imagine how a honey bee could remove mites and a wide range of particles. It seems that the process of honey bee grooming produces lots of body hair and other particulates that find their way into debris (shown in figure 4.1.8).

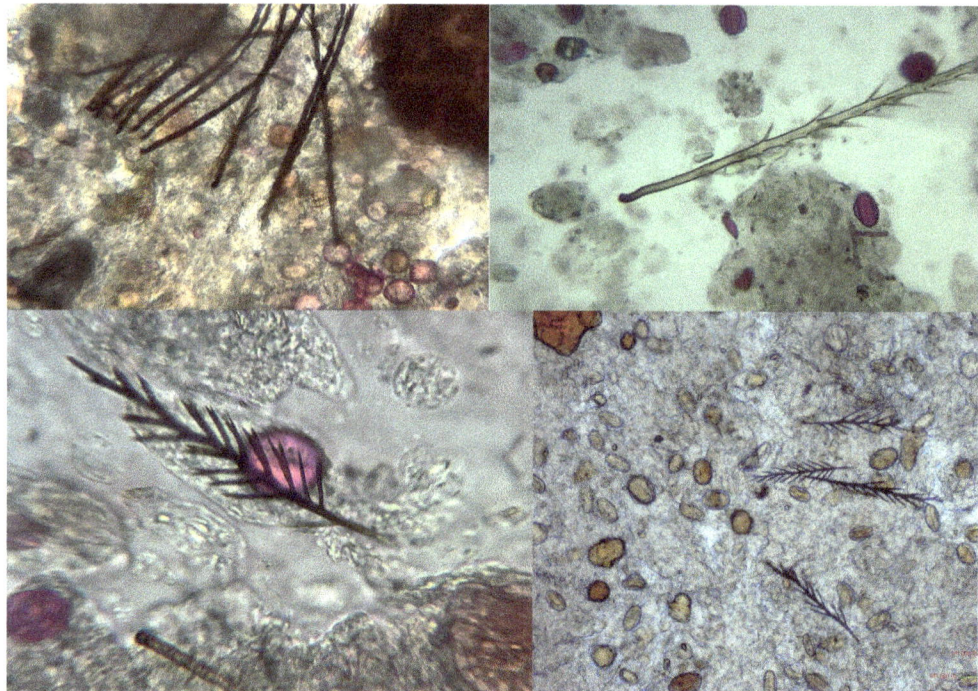

Figure 4.1.8 Bee hairs

A plausible and straightforward explanation for body hair in the debris is hygienic behaviour. However, as I looked down the microscope and counted hairs, something more akin to a crime scene was unfolding. Body parts of all shapes and sizes, claws, legs, antennae, exoskeleton fragments, stingers and so on. Sometimes these were found alone on the waste board and sometimes entombed in wax. As I approached the end of January, I wished that I had made the time to note these unexpected bee parts. I did however record a few interesting finds. This one became a fascinating study of honey bee anatomy (see figure 4.1.9)

It shows the adaptations of a bee's foot, especially the spongy lobes at the end of each foot that helps a bee stick to surfaces that are too hard for the claws to grasp. I'd never really thought about the mechanics of a bee's foot.

While looking at the claws, I thought that's a strange place for hair. On closer examination, the claw is closed tight, grabbing a 'clawful' of bee hair (look closely at figure 4.1.9). It seems like a grooming bee has pulled these hairs from another bee before having its own foot amputated - is this a story of grooming, collateral damage while grooming or, is there another story to be told? Feet are a common find - I'd never thought about looking to see what is in the two clawed toes at the end of each foot.

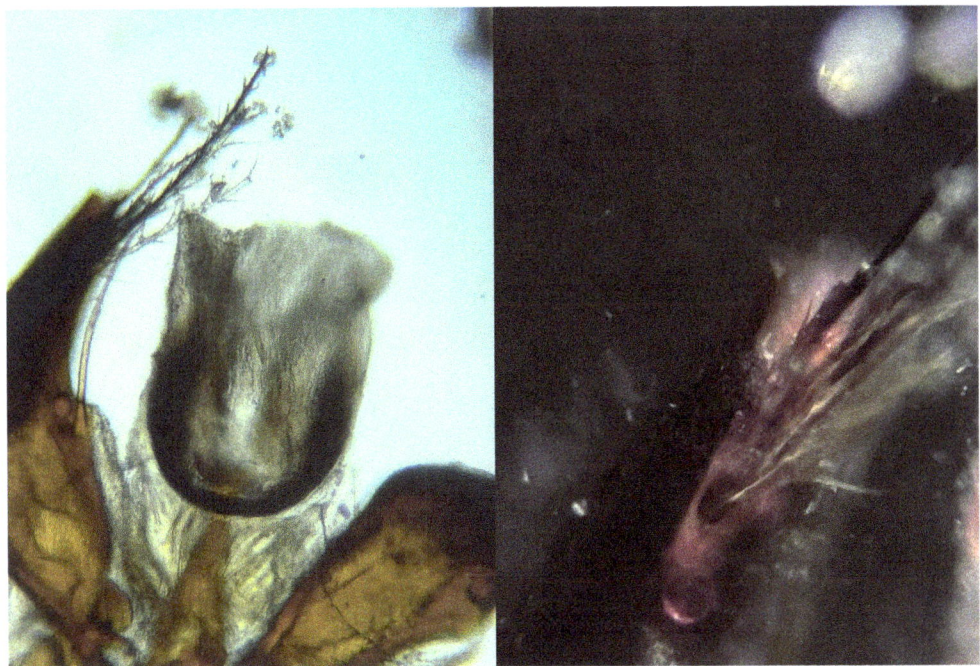

Figure 4.1.9 Spongy feet and grabbing claws

I remember learning that mortuary worker bees take dead bees and remove them from the hive, hygienic behaviour to reduce the likelihood of disease. A nice solution and narrative about bee cleanliness and organisation. I started to wonder if these bee parts are really the debris created by mortuary bees, are bees breaking up bees to easily remove them from the hive entrance or, if there is a different story waiting to be told? One to watch and ponder in future months. I must remember to take some pictures of body parts and do some background reading about this.

Well, that's the samples for January completed. Time now to put leftover debris in the freezer (just in case it is needed later), wash microscope slides and plan for February. How will things change? Will I be dancing for gold?

4.2 February

Mites and textiles in the debris

Honey bee debris collected between 1st and 28th February

Minimum temperature -6°C

Maximum temperature 13°C

Mean temperature 6°C

Compared with the previous 28 days there has been an almost 7% increase in the total debris produced by this colony. This increase of less than 0.5g is difficult to see on graph 3.1 due to the scale used. However, with careful observation it is possible to see this increase in colony activity by comparing the two debris profile photographs. You would however need to place these two pictures next to each other to spot this small yet important change. Highlighting the importance of saving images rather than trying to remember the debris pattern.

The two pictures show an expansion of activity towards a northerly aspect of the hive. It is also interesting to note that there are many more wax structures away from the heart of the cluster. For now, I will call these wax shards. It is possible to see how many of these would have been part of a wax hexagon structure. Is this some kind of comb re-engineering? Cutting and re-fabrication of cells & for what purpose, accessing nutrients or preparing the comb for the coming months? The bees are investing valuable energy in a process that results in these shards falling onto the inspection board. There is also a droplet of sweet tasting liquid (yes, I tried it, although wouldn't recommend it and plan to buy a sugar testing kit for any future droplets, should they occur) in quadrat C4

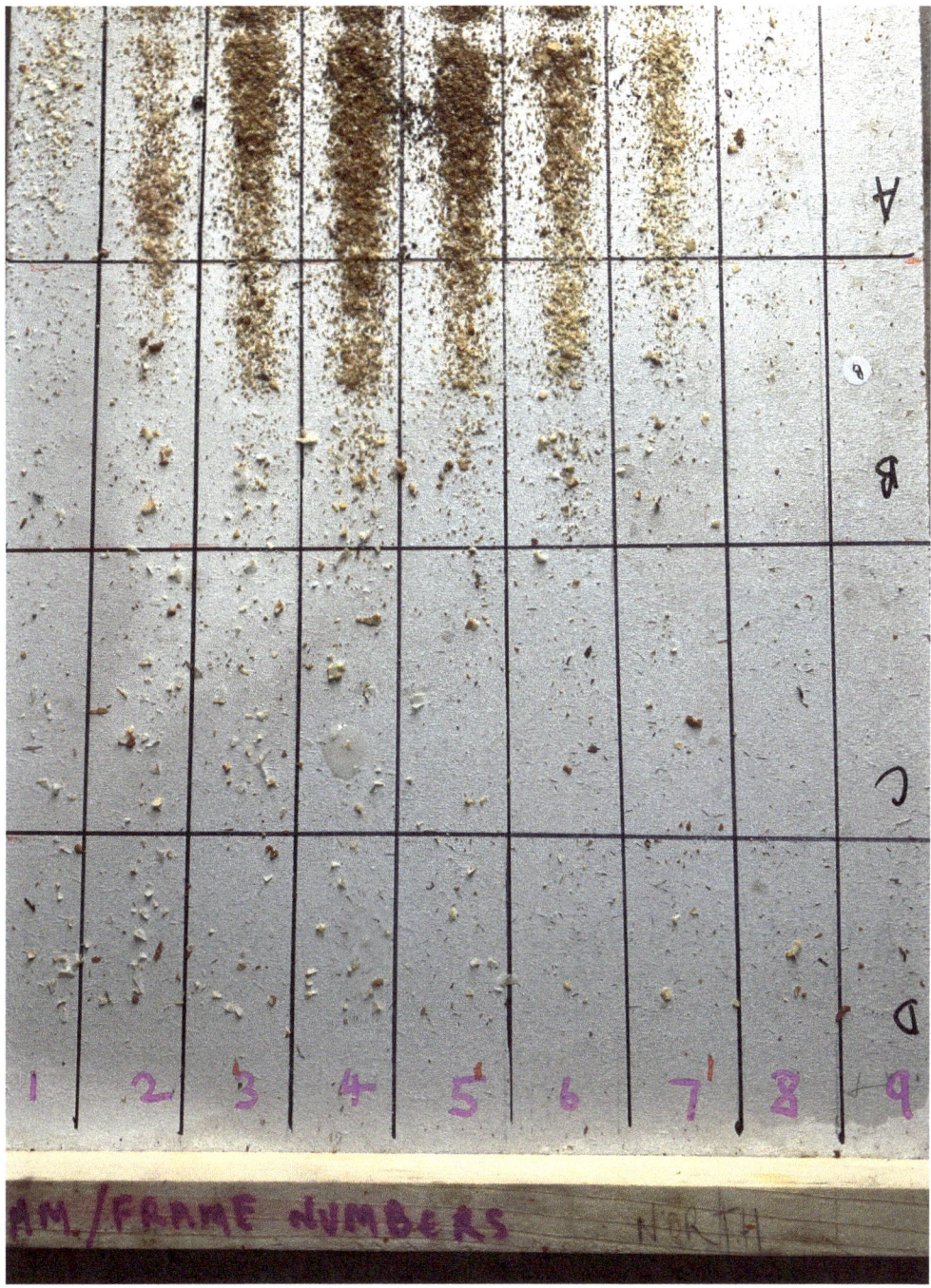

Figure 4.2.1 February debris

New pollens for a new month

Figure 4.2.2 shows honey bees foraging on hellebore, spring heather and plum in February. To watch honey bees enthusiastically working these flowers warms the heart, especially knowing that we planted these flowering plants and to see some of the fruits of our labour. These early foragers are taking advantage of the increase in day length and mean temperatures. It still feels very dark and cold to me, but the changes are enough to bring about the first signs of spring in the debris. Pollen from these three flowers were the most common finds in the debris (see figure 4.2.3).

Figure 4.2.2 Fresh forage in February

Other pollen found in the waste includes spiky Asteraceae pollen of which we have lots. The torpedo shaped pollen is from field beans planted by the farmer next door. Below the field bean pollen is hellebore. The phacelia pollen (bottom right) is most likely from a cover crop grown about a mile away. These 'crop' pollen grains must have been collected and stored and preserved by the bees more than six months ago.

This 'fresh' pollen has been either accidentally dropped on the way to a storage cell or during pollen storage itself. Bees perform gymnastics to pack pollen into cells. The forager enters a cell backwards, with her rear legs and pollen loads entering the cell first. This worker hangs onto the upper cell wall and uses her other legs to brush off the pollen from its rear legs into the cell. Leg cleaning and pollen pushing is repeated several times until the legs are cleaned. This compacted pollen is then covered with a thin layer of honey and wax and the process of fermentation will preserve this nutrient-rich food source for the bee colony (Siefert et al. 2021). It is easy to imagine how pollen could be dropped, or other debris created, during the physical activity of storing pollen.

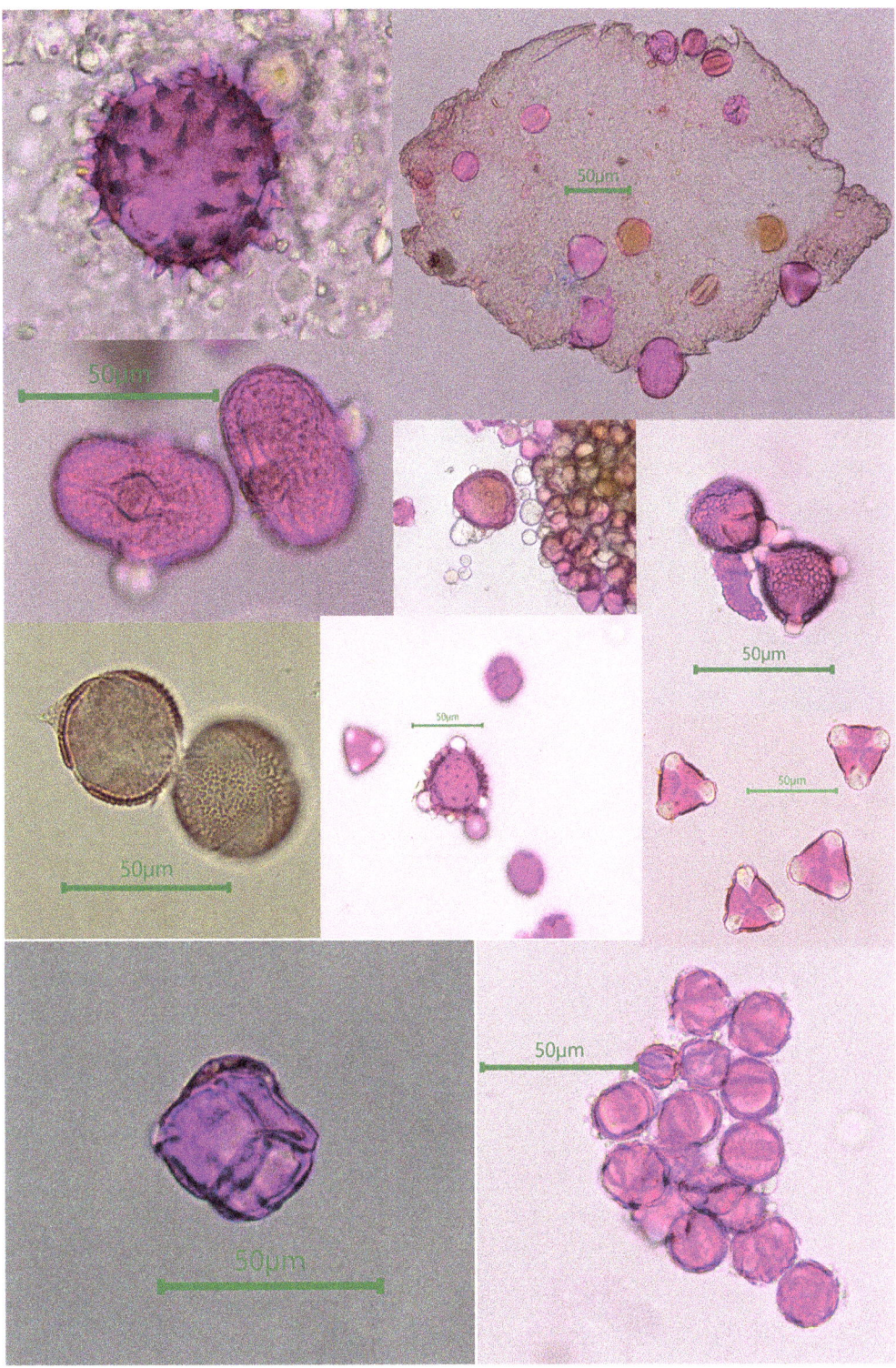

Figure 4.2.3 February pollen

Exoskeleton parts

Last month, I resolved to pay more attention to the body parts turning up in the waste. In the February debris exoskeleton parts could be found in two different forms.

- 'Fresh and naked' exoskeleton fragments
- Exoskeleton fragments are found entombed in an amalgam of propolis and wax

There were hundreds of these fresh fragments in the debris. Figure 4.2.4 would make a good 'guess that part' quiz for beekeepers. I can see a sheath from a stinger, three ocelli, the sockets for two antennae, compound eyes, and the leg of a mite are mixed up in these pictures too. Three things that surprised me about these images are, i) that all the parts seem to be hard outer bee parts with no bee innards visible, ii) the baldness of exoskeleton parts, and iii) the pattern of the torn edges of exoskeleton parts. Could these clues help explain what happened to these bees?

Figure 4.2.5 helps to illustrate these common features. On a living bee the head and the eyes are covered in mechanosensory hairs. This piece of head is however completely bald. Also the crack in the skull hints at the strength and cutting power of honey bee mandibles.

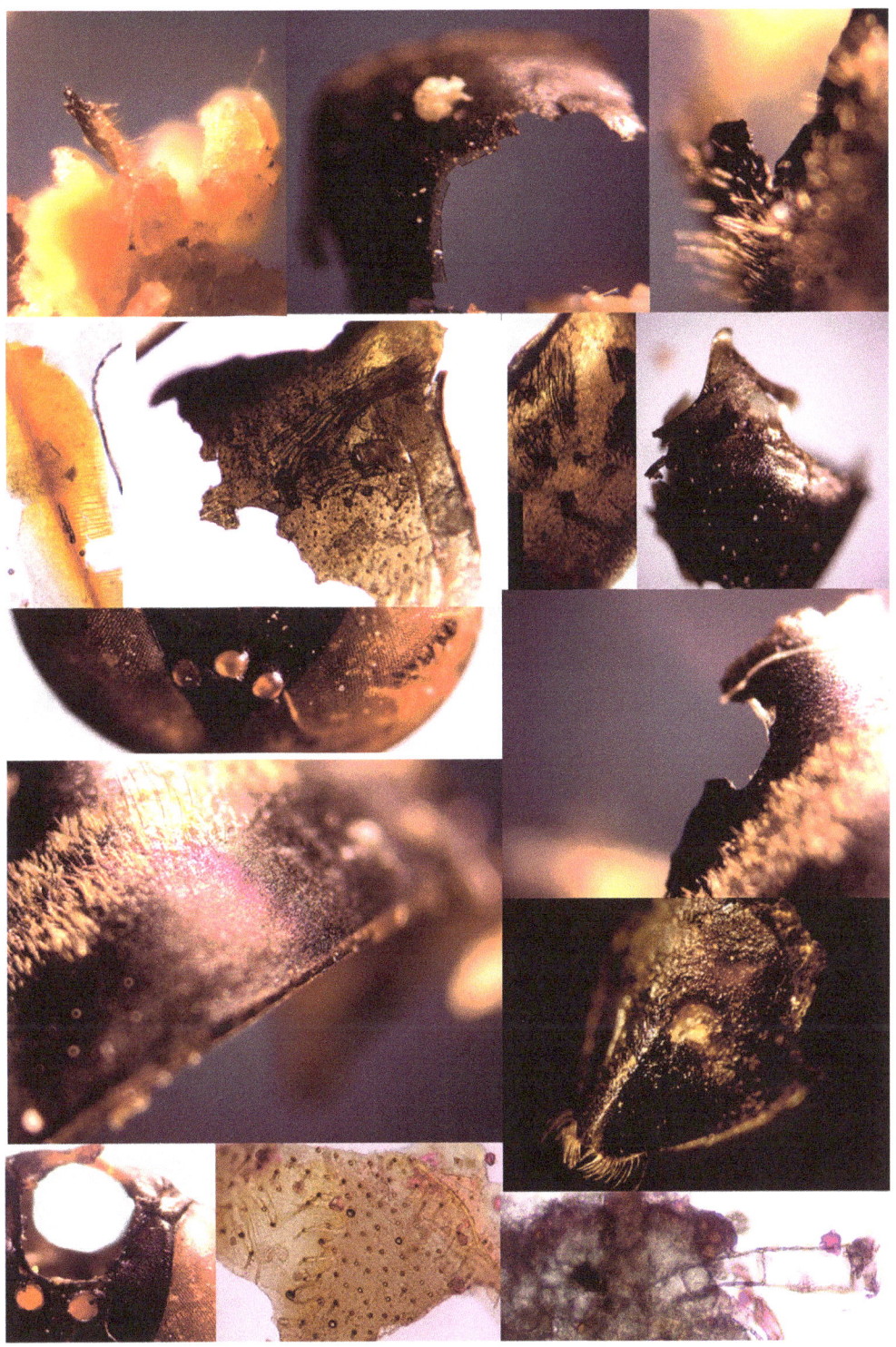

Figure 4.2.4 shows just a small sample of some of the 'fresh' exoskeleton parts visible with the naked eye on the waste board. Figure 4.2.4 Hard body parts

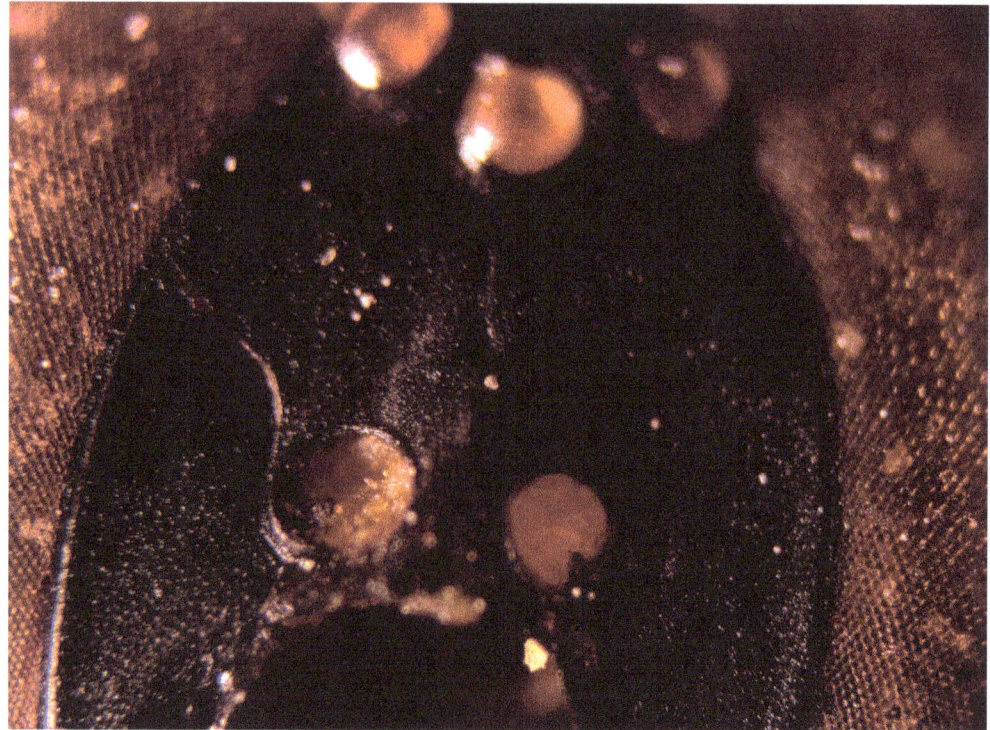

Figure 4.2.5 Forehead of a bee

The forehead (frons) of a honey bee is very tough indeed. Not surprising as it provides structural support and protection for a complex sensory system. At the top of the image are three small ocelli (eyes). Below are the two holes for the socket joints that enable the antennae to move in all directions. On the sides of the image are the edges of two compound eyes. In a living bee all these parts would be covered in hair.

Bee pellets

Figure 4.2.6 shows the second form of exoskeleton parts found in the debris. These are much more numerous. These fragments are found entombed in an amalgam of propolis and wax. Dissecting these tiny waxy bee pellets is like dissecting owl pellets, a favourite lesson at school. Barn owls cannot digest the fur and bones of the small mammals which they eat. The owl swallows their prey whole, absorbs the nutrients and then a few hours later, coughs everything up that is of no use to them.

Something different is happening with honey bees' wax pellets, but equally fascinating. They are similar due to being distinct pellets containing all kinds of debris and each has a story to tell. The picture shows exoskeleton fragments, fibres and hairs present in these pellets.

Figure 4.2.6 Bee pellets - entombed exoskeleton fragments

Fibres - warp and weft

In January there was an estimated 15738 fibres in the debris. In February this value increased to 16773. Figure 4.2.7 shows just a few of these fibre finds. Most of these were found in waxy lumps after dissolving the wax using a solvent to reveal the fibres within. It's not always easy to say whether these are natural or man-made fibres. However, the picture of warp and weft textile (see top right) embedded in the wax left me in no doubt that there are man-made fibres in the debris of my hive. Also, the several red lipstick-coloured lumps found are difficult to explain as natural objects. It seems that we should definitely be talking about man-made fibres in honey bee colonies. Figure 4.2.8 helps to make this point - what is the grooming burden for bees with fibres wrapped around their legs?

More evidence from February demonstrates that some of the waxy lumps found on the inspection board are a product of grooming and waste processing. These waxy bee pellets are a composite of many parts, containing; bee parts, mite parts, bee hairs, fibres, pollens, and fungal structures.

Mites

An exciting find this month has been watching the many tiny mites (see figure 4.2.9) go about their business. I am no mite expert but using the identification key provided by Bee Mite ID (USDA, 2016) my best first effort suggested that this mite belonged to the Saproglyphus genus. However further study suggests it is more likely to be Carpoglyphus lactis. This uncertainty is due to difficulties using morphological features of a very small moving mite for identification. DNA testing would be helpful to confirm identification.

Both species are whitish with an oval body about 0.4 mm long. Their mouth parts are pointed and each leg ends in a small claw. Both species are known to live and feed on decomposing organic materials and to be associated with honey bee colonies. While mites are commonly called pollen mites, the term 'pollen' is troublesome, because it suggests that this mite lives out its life on pollen. In this study, these mites have been found in waxy lumps with what looks like fungal growth on the surface. Perhaps 'debris' mites would be a more accurate description. Whether these mites are harmful to bees is a question that will be discussed later in this study.

Bee Mite ID states that "Carpoglyphus lactis is a relatively common species in beehives, occurring in the debris on the bottom boards of beehives, honeycombs, dead bees, honey, and especially on the bee bread (bee pollen with added honey and bee secretions that is stored in brood cells). This mite can penetrate the brood cells and make burrows in the stored pollen, consuming it and causing the pollen and the debris to spill from the cells. The infested bee bread, mixed with large quantities of dead and live mites, turns to a golden-brown or yellow powdery material covering honeycomb and bottom boards of beehive".

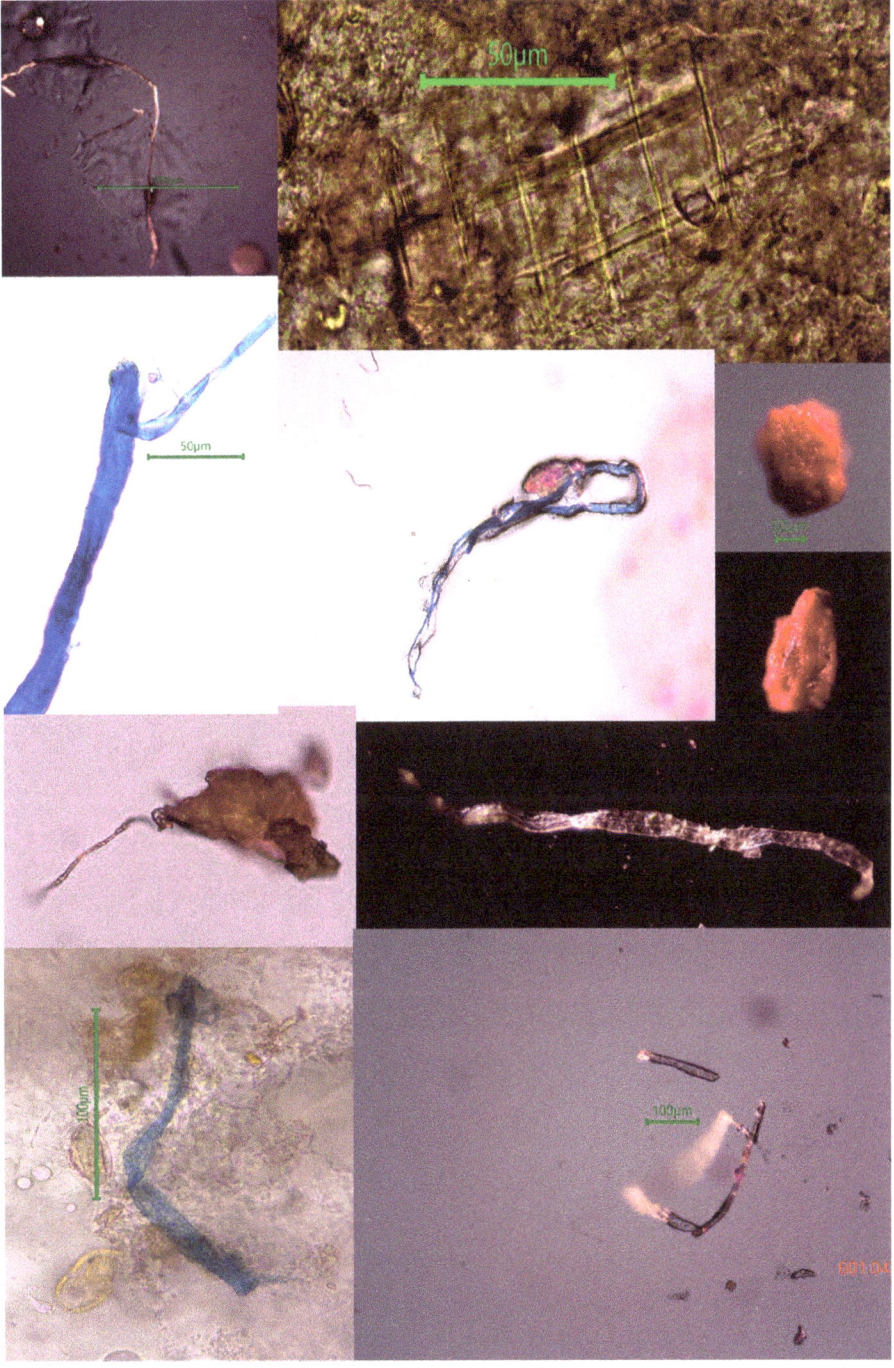

Figure 4.2.7 Fibres found in the debris

Figure 4.2.8 Fibre attached to an amputated claw found in the debris

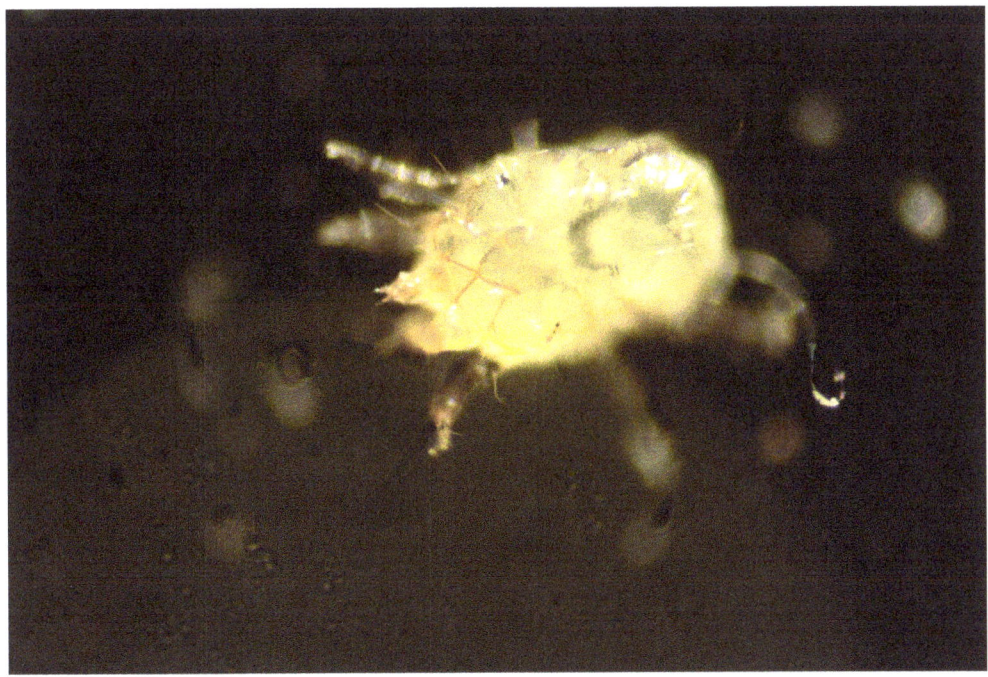

Figure 4.2.9 Bee debris mites

I wanted to investigate further, so I prepared for a mite hunt. It did not take long to discover one and also unexpectedly some tiny eggs. The mite eggs can be seen in the dark crevice of the waxy lump with what looks like mycelial growth on the surface. An exciting find, suggesting these mites may well live out their lifecycle on debris.

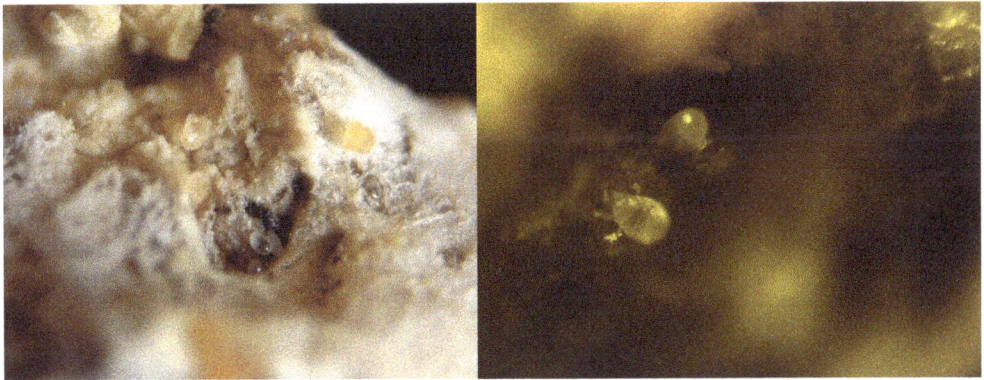

Figure 4.2.10 Mite eggs in the debris

I wanted to see if it was possible to watch an actual egg hatch in wax, showing that the wax is indeed the place where the mite reproduces. To help answer this question, I kept the same mite eggs on the same waxy lump and watched at 4-hour intervals using a microscope magnification of 40 x and 100x. I got lucky and 24 hours later I managed to capture actual video footage of a mite emerging from its egg. Figure 4.2.11 shows the newly emerged mite, with only 3 pairs of legs. Later, in its lifecycle it will have 4 pairs of legs. Two weeks later, I checked in again on the same wax lump. All the eggs had gone and there was a young adult mite, in the same crevice nest site as the mite eggs. Also, the size of the crevice was much larger. The waxy lump of debris had been broken down, suggesting that these mites may well have a role as decomposers making a living in the debris.

Figure 4.2.11 An emerging mite

(Bell, 2014) states that the mite Carpoglyphus lactis is a common pest in food environments utilising microenvironments of moderate temperature and raised humidity. The same mite is well known amongst growers of dried fruit as a pest that can cause major problems with food spoilage.

"The life cycle includes a brief larval stage typically followed by three nymphal stages prior to the reproductive adult stage...with only 14 days being needed to complete development under optimal conditions and with a single female being able to produce 555-600 eggs. Eggs are cold-tolerant and in some species, development can proceed down to 5°C, but in all species low humidities prevent development. (Bell 2014).

It is possible that humidity is a key factor in the control of these mites. In the world of dried fruit, prunes can be kept mite-free when the water content is below 17% and they are kept in a well-ventilated room (Pestium, 2024). It is interesting to note that capped honey will normally have a similar moisture content of approximately 18%, while the moisture content of nectar is much higher at around 80%. Like with prunes the moisture content of honey is a quality indicator that determines the ability to avoid spoilage. It seems plausible to say that any beekeeping practice that raises the humidity nearer to optimal values risks an increased mite population and spoilage.

4.3 March

Spring growth

Honey bee debris collected between 1st and 28th March

Minimum temperature -3°C

Maximum temperature 14°C

Mean temperature 6°C

In March there has been a 68% increase in the weight of debris deposited on the inspection screen compared to the previous month. It seems fair to assume that this represents a similar increase in colony activity. This view is supported by the area and composition of debris on the inspection screen (see figure 4.3.1) and the heat maps (figure 3.2). Both show the colony expanding its range of activity from what was the core of the nest. The volume occupied by the brood is expanding and resulting in debris created by nurse bees caring for the young brood following egg laying. In previous years, I got my kicks from watching activity at the hive entrance and gleaning as much information as possible. Sometimes, I would even take a flask, deckchair, and notebook to record comings and goings. I would look out for bee behaviour and tightly packed pollen lumps on the hind legs of bees. This year I am learning so much more by looking at the bottom of the hive. I wish that I had a record of the debris from all the colonies that I have visited over twelve years. Had I done that then I would have a debris record for more than 100 colonies - a record of successes, failures, and everything in between.

Figure 4.3.1 March debris

Foragers returning with pollen are commonly believed to be a sign that the colony is queen-right. It is these observations, along with a spell of good weather that helps beekeepers to plan the ideal time to carry out the first inspection of the year. This year however, I'm spending much more time looking at the debris collected at the bottom of the hive. It is giving me the same information and much more than can be observed at the entrance and has the advantage of being able to be made regardless of the weather.

Pollen grains collected from the debris

Whole pollen grains could easily be spotted amongst the debris (see figure 4.3.2). These multi-coloured lumps of pollen contain an essential source of protein, fats, vitamins, minerals, and carbohydrates that are needed for the growth and development of a healthy colony. These compressed pollen lumps are also transport vectors for mites, fungal spores, plant structures, agrochemicals and many other microorganisms. It seems logical to assume that all of these will in some way end up being deposited in the debris. In the hive it is likely that these will affect the colony in different ways. Sometimes these will be a burden and on other occasions that they may well be beneficial to the colony.

Figure 4.3.2 Compressed pollen lumps

Microscopy shows that the debris contained pollen is from flowering currant, wild cherry, spring heather (top left), willow (top right), and daffodil (see figure 4.3.3). I like the boat-shaped structure of daffodil pollen. It is interesting that daffodil pollen appears in such large numbers. Over the years, I have heard from many sources that bees don't feed on daffodils, or that daffodils are a food of last resort. The sheer amount of daffodil pollen would suggest that both statements are untrue. The most common pollen grains in March are willow and wild cherry, with a few spiky Asteraceae pollen structures and field bean grains to be found from last year's forage (not included in this pic). The pollen structure second down on the left looks like lavender with 6 pores and furrows.

The composition of this month's debris is very different to preceding months, so much so, that I started to doubt the trends. 'Result-anxiety' set in, so I ran the whole experiment again and guess what? The same trends repeated themselves.

While the weight of the debris increased, the number of bee hairs per gram and chalkbrood cysts per gram of debris decreased (see chapter 3). These reductions in spore and hair 'density' have perhaps been caused by a 'dilution' of these parts by new kinds of activity. The spring expansion of the colony is changing the composition of the debris. One of these changes is the increased number of beeswax structures. While some of these structures can be seen using the naked eye or an enlarged photograph, a microscope adds new detail to tell a story that connects the micro to a macro understanding of bee nest architecture and the wider environment.

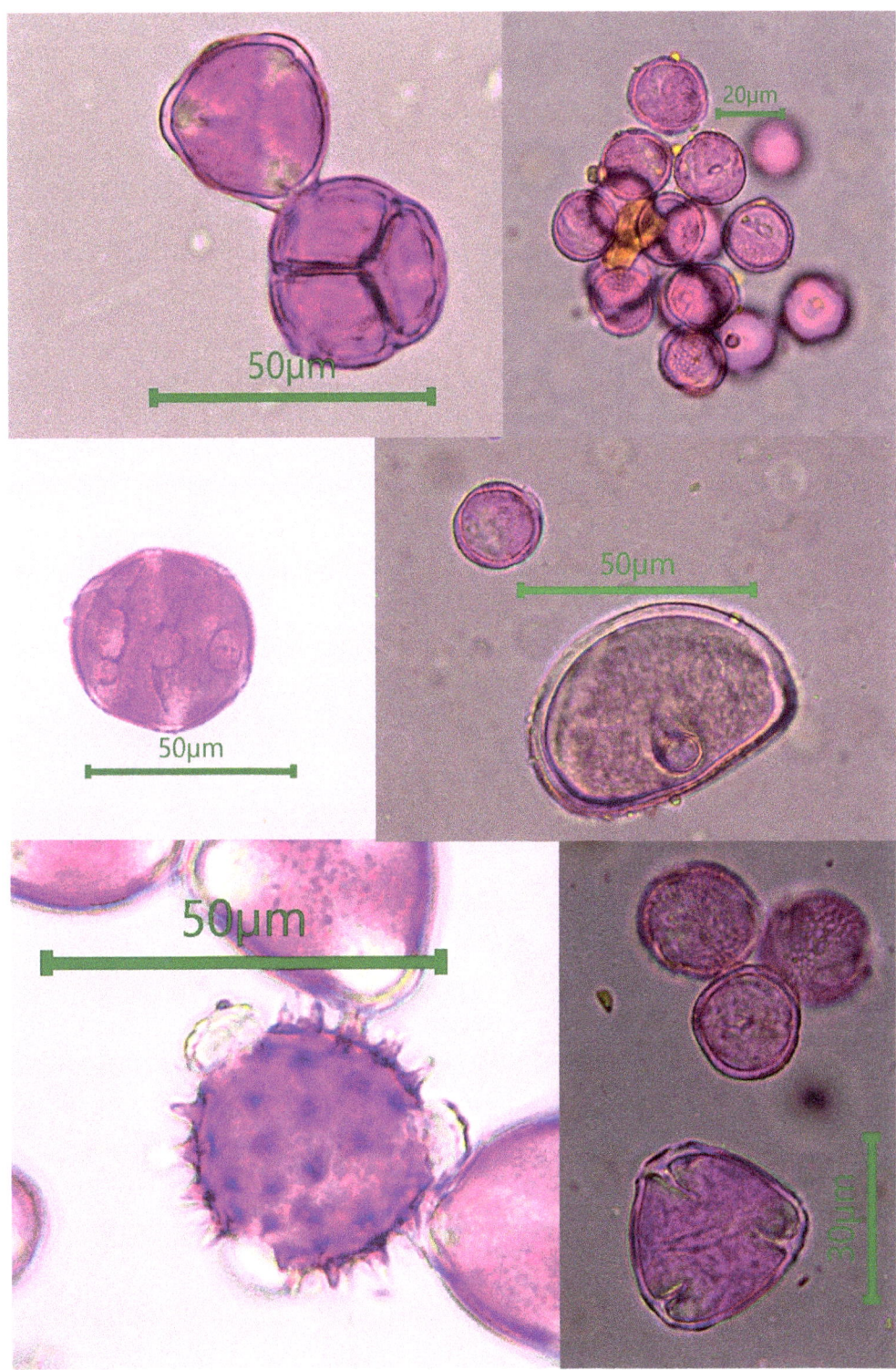

Figure 4.3.3 March pollen

Beeswax in the debris

Beeswax is found in four different forms.

Figure 4.3.4 shows two of these forms. First are the transparent wax scales which are produced by 4 pairs of abdominal glands. A liquid when secreted, it then becomes unique solid pearly white wax scales after cooling. These scales are then transported around the nest by bees using their fine leg hairs. It is possible to see the puncture marks in these wax scales, probably caused by leg hairs during transportation.

Figure 4.3.4 also shows a lump of stained wax which looks like it has been processed from old wax.

Some of these fresh wax scales have been 'welded' together at a midrib (see figure 4.3.6). It seems that these scales would then be used directly in comb construction. To think that this beeswax has been manufactured from local plant sugars is amazing, and then to consider the fabrication and welding skills needed for nest building is incredible.

Figure 4.3.5 shows freshly chewed wax lumps that have been mixed with saliva, processing ready for some kind for comb engineering.

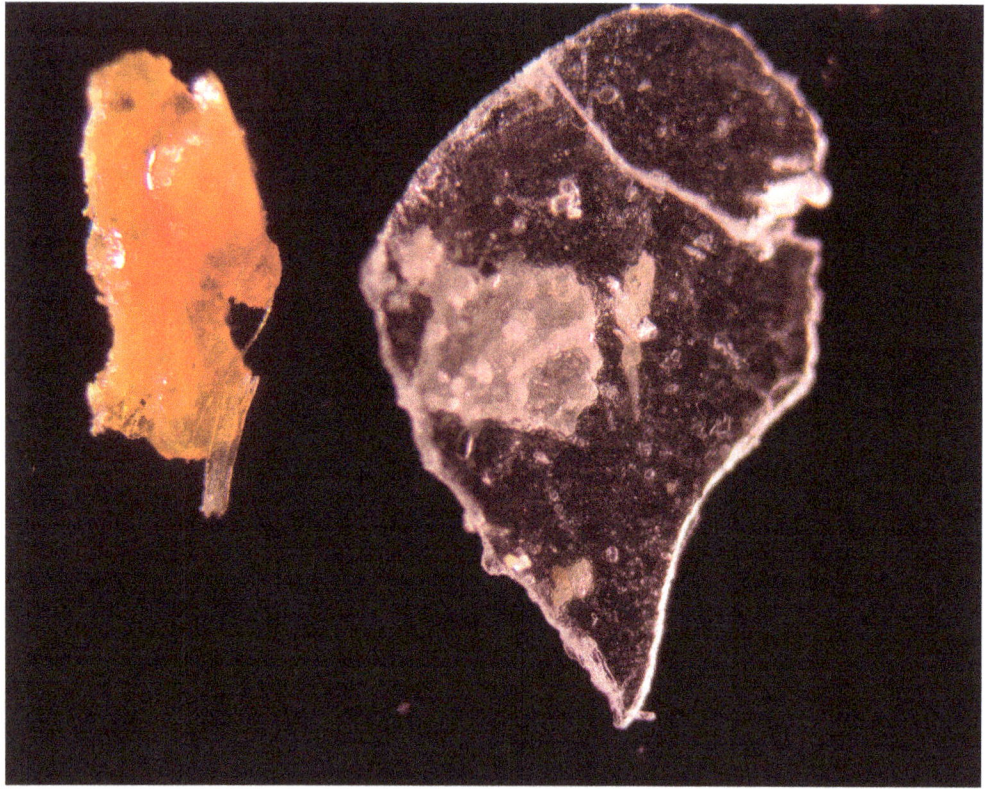

Figure 4.3.4 Wax scales and stained chewed wax

Figure 4.3.5 Chewed virgin wax

Figure 4.3.6 Wax welded at the midrib

Bees manufacture wax when they need it and when resources are available. New wax is a helpful indicator showing when there is a nectar flow, brood rearing, the presence of a queen, the presence of pollen, a young workforce, and the suitable brood nest temperature. Finding fresh wax says a great deal about a colony and the nest environment.

Varroa - getting to know the enemy

I have been surprised not to find any mites on the inspection board. Figure 3.5 shows a high population of mites in the autumn of 2023 and none in the spring of 2024. It seems fair to assume that the reduction in mites is a result of an oxalic acid treatment given in late October 2023.

Varroa mites are a serious pest for all beekeepers and an even bigger pest for honey bees. While they are wingless, eyeless, and unable to fly across the gap between brood frames, the varroa mite has been remarkably successful at infecting most honey bee colonies in the UK. No other parasite has had a comparable impact on honey bees (Traynor, et al., 2020). They need to be taken seriously.

The life cycle of the mite consists of two phases.

Varroa reproduce within brood cells and feed on developing pupae. This is called the reproductive stage. Recent research demonstrates that they feed on pupal haemolymph during this reproductive phase. The dispersal stage is when the mite attaches itself to the adult bee and feeds on the fat body tissue (Han, et al., 2024). The honey bee becomes a vector transporting the mite in and outside the hive providing many opportunities for the mite to attach itself to other bees.

The varroa mite is also a well-known carrier of many viruses that can also cause serious harm to honey bee colonies. Honey bees are harmed by this type of parasitism in many ways, including a reduction in weight and lifespan of emerging bees. Beekeepers are still learning how to live with and control varroa, as are the bees. The species of varroa that affects UK bees is a relatively recent arrival and has not yet achieved a host-parasite equilibrium. Three kinds of adaptive behaviour amongst honey bees have been observed. It is not difficult to imagine that each of these behaviours will produce its own debris trail, a mixture of whole varroa, varroa with bite marks, varroa parts and infected pupae. While it may be impossible to know the history of each mite in the debris, over time findings may give a useful indication of these adaptive behaviours.

1. Grooming where adult bees use their mandibles and legs to physically pull the mite off the body off the honey bee.
2. Hygienic behaviour when infected pupae are removed from cells. It is believed that olfactory cues are detected from damaged broods and that these chemical signatures diffuse through the brood cap and in turn stimulate other adult bees

to remove the infected pupae.
3. Biting where adult bees use their mandibles to damage and/or kill mites. (Rosenkranz, 2020) found considerable variation between colonies in their ability to immobilise varroa by biting.

Figure 4.3.7 shows two mites collected in November 2023. One of these mites is whole and the other has missing legs and damage to the exoskeleton.

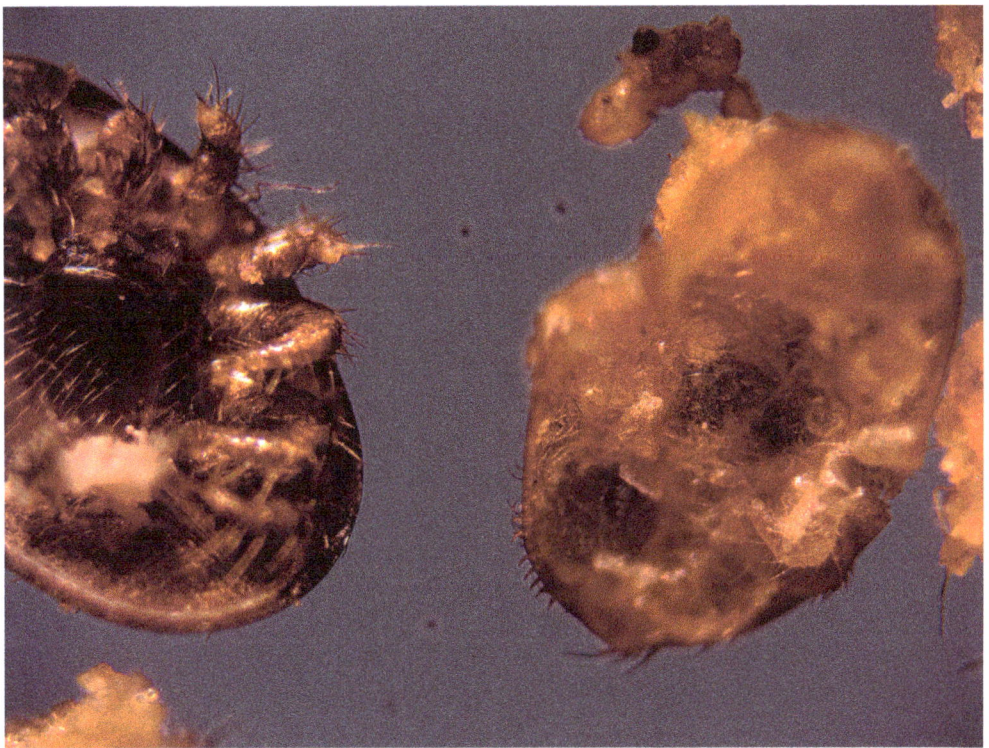

Figure 4.3.7 healthy v damaged mite

Monitoring mite levels.

There are several ways to monitor mite populations, each one has pros and cons. The simplest way is to count the whole mites that drop onto the inspection board (in later months I found that microscopy can reveal many more mites than those visible on the inspection board). This number gives an approximation of the number of mites in the colony. A higher number of mites indicates a greater problem than a smaller number.

I have been counting/recording mites on the inspection board since September 2023. Figure 3.5 shows the number of mites found on the waste board in a 28-day period for each month. By the end of this project the graph will show a 16-month record of the mite drop for this colony. These mites represent either natural mortality or deaths caused by varroa treatments.

Any figure above a daily mite drop of 30 is considered serious and immediate treatment recommended. Below 10 mites per day is considered acceptable with no recommendation to treat, and a daily drop of between 10 and 30 the advice is to treat soon (Stainton 2023). I am pleased that for the past seven months the mite population has mostly seemed under control. The highest mite drop occurred in November with a total mite mortality of 335. This higher death rate is likely to have been caused by an oxalic acid trickle treatment. The reason for treating this colony was to reduce the winter mite number to as low a number as possible. Adult mites are known to over winter, attached to bees for up to 6 months. Like winter bees, they have a much longer lifespan in the winter. It is these fertile mites in their dispersal stage (formerly known as phoretic) that pose a risk for the coming season.

The average effectiveness of a trickle treatment of oxalic acid is quoted above to be 91.4%, meaning that it kills most varroa in the colony. This efficacy number is an average value from 10 different UK studies. It doesn't, however, kill all mites.

If this treatment has killed 91.4% of the mites, then that also means that 8.6% of the population have survived. Using the results shown in Figure 3.5, a survival rate of 8.6% suggests that 32 mites may be overwintering, attached to workers and waiting for new bee brood in the spring. A small but significant number that, given the right conditions, can quickly overrun a colony. These varroa will get into the brood cells again as soon as the queen starts laying after the winter Solstice.

A thought experiment to understand the risk for 2024

It is known that on average each mother mite will produce 1.5 mites in worker brood and 2.5 mites in drone brood per brood cycle. To keep the maths simple, let's say that each mother produces 2 mites per brood cycle. This will double the mite population every 21 or 24 days (depending on whether it is worker or drone brood). Using a best guess that 32 mites are over-wintering in this colony, after one brood cycle there will be 64 mites, after 2 brood cycles 128 mites, 256, 512, 1024 and so on. With optimum conditions for reproduction and assuming no limiting factors on growth there will be more than the value that The National Bee Unit (UK) says puts a colony at serious risk (The National Bee Unit, 2024). This estimate has the potential to be an underestimation as it does not include the mites that are carried by adult bees. Neither does it consider the hygienic behaviour of bees or any colony ability to self-regulate the mite population. This crude calculation is an attempt to use the debris to monitor the varroa population in this colony. As the months progress it will be interesting to compare these predictions with the monthly mite drops.

When I wrote this paragraph, I fully expected to see mites by April but was delighted that I had to wait until August before evidence of varroa was observed in the debris.

Mites and fungi

As in previous months I haven't been counting 'pollen' mites, but there are many. I can see that they are very good at picking up all kinds of debris on their feet as they walk. I now think of them as mini vacuum cleaners, collecting and consuming all manner of organic debris. Several could be seen walking up and down an amputated hairy bee leg. It's nice to write about a mite that may well be beneficial for the colony, rather than varroa, which grabs all the headlines. There were also many alternaria fungi in the debris (see figure 4.3.8). I have regretted not counting the mites and alternaria to monitor their populations over time.

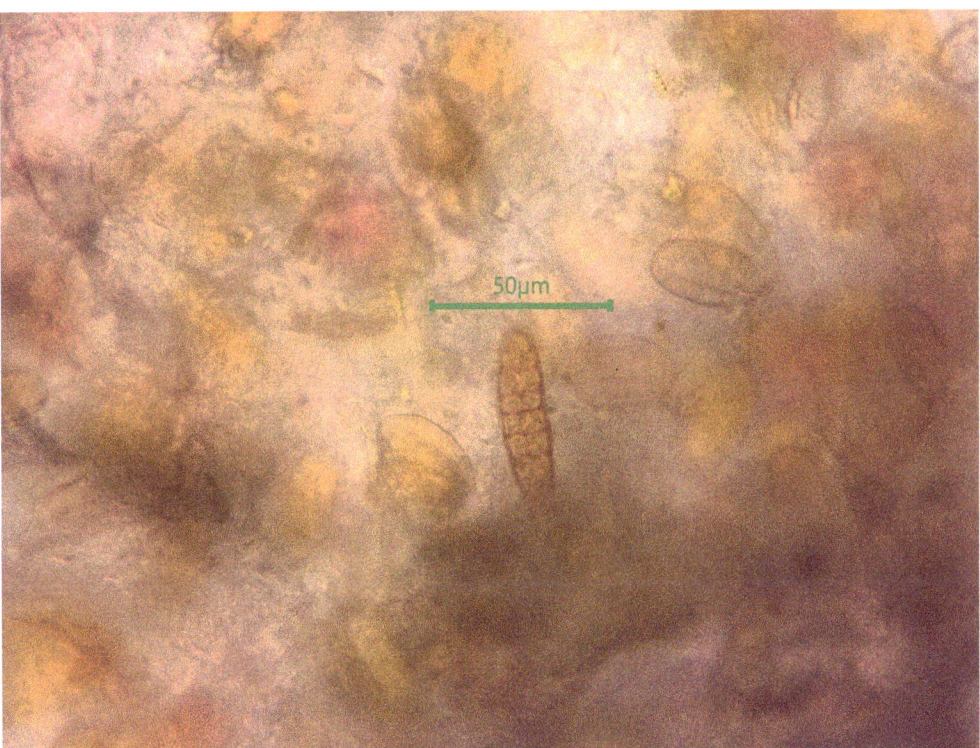

Figure 4.3.8 Alternaria fungi

4.4 April

Rapid growth

Using a photo record as a proxy indicator for colony activity

Honey bee debris collected between 1st and 28th April

Minimum temperature -1°C

Maximum temperature 18°C

Mean temperature 8°C

I don't need to open this hive to see where and how the brood nest is expanding. The photorecord of debris deposition is showing this (see figure 4.4.1). The results shown in chapter 3 show that there has been more than a fivefold increase in debris in the first quarter of this year - a proxy indicator for colony activity. Also, the changing composition of the debris, the whole brood caps, pollen grains, increasing number of wax structures as described in the previous month and a lack of varroa mites, suggest that there is a young workforce looking after a growing brood nest, with adequate nutrition and a laying queen.

Regular hive inspections have become the cornerstone of modern beekeeping. A top-down model of beekeeping that only became possible due to the discovery of bee space and the invention of top opening hives with removable frames.

L.L Langstroth describes how this revolution in beekeeping practice has firmly placed humans in charge of honey bees.

"... the chief peculiarity in my hive was the facility with which they could be removed without enraging the bees...I could dispense with natural swarming, and yet multiply colonies with greater rapidity and certainty than by the common methods...feeble colonies could be strengthened, and those which had lost their queen furnished with the means of obtaining another...If I suspected that anything was wrong with a hive, I could quickly ascertain it true condition and apply the proper remedies" (Langstroth, 1880)

The regular removal of frames and their 'shuffling' soon became the husbandry industry standard and today remains the keystone practice that enables the beekeeper to detect pests and diseases, assess vigour (the physical health and strength of a colony), for queen rearing and to plan interventions like creating a nucleus or feeding.

In this study I am leaning towards a more bottom-up approach by trying to decipher the information contained in the debris and using this information to guide my own practice. This is a move away from opening boxes and ripping apart the nest to make judgements. It's a different way of doing things that challenges many of my beliefs and assumptions about hive inspections. I am not advocating that people stop doing traditional hive inspections, far from it. It's just that studying debris can help us to be better informed before deciding what to do next.

Figure 4.4.1 April debris

Like all beekeepers, I just love the close interaction with honey bees. There is something almost magical about the experience that only a beekeeper can understand. The experience of carrying out some kind of intervention without overstaying my welcome is priceless. It's almost May and I have not inspected this colony, and quite frankly I am missing the experience. This itch along with a desire to 'verify' the conclusions being made from studying debris were the only prompts needed to carry out a quick inspection. After climbing into a bee suit and removing a few frames it was possible to see that the shape of the brood nest was an approximate duplicate of the darker parts of the debris screen (see figure 4.4.2). Also, the emerging worker population and sealed brood were located directly above the wax cappings found in the debris. Frames removed from the centre of the brood area showed no queen cells. This inspection and the debris indicated that the population would continue to expand at a similar rate. It was plain to see that the expanding brood population would soon run out of space. In the absence of any queen cells a decision was made to remove the queen excluder and give the queen extra space and try to reduce the swarming impulse.

Brood Caps - a double skin structure.

Brood caps differ from honey caps in ways that relate to their function. A honey cap is built almost entirely from wax. Its construction prevents the reabsorption of water that would spoil food stores by fermentation. A brood cap has a very different function, with a structure to match. A highly permeable capping allows the diffusion of gases for respiration. The brood cap also provides a cue for the larvae to spin a silk cocoon for the metamorphosis of a larva into a pupa. Covering the glossy silk cocoon with a wax capping is known to be beneficial for optimal brood development by buffering the humidity and temperature conditions. In experimental conditions it has been shown to be possible to raise brood without wax cappings by artificially controlling the temperature and humidity (Kubasek, et al., 2022). Under a microscope, a brood cap looks more like a double layered textile. The study of nibbled cap edges is useful to learn about the structure and imagine bees emerging from the safety of their cocoons. Figure 4.4.3 shows the outer layer of each cap is a thin layer of processed wax mixed with propolis. Figure 4.4.4 shows the inner layer that consists of thin silk fibres which is laid down from beneath. In the debris, only brood caps were found with no honey caps falling on the inspection board. I believe that this is due to how brood caps are 'opened' by emerging bees. A process that seems similar how we open a can of beans, resulting in an intact lid/cap with a 'jaggy' edge. Only a small percentage of the total number of brood caps end up in the debris and none have been seen being removed bees at the hive entrance, so it seems most likely that these resources are recycled and repurposed by the colony.

Figure 4.4.2 Brood from above the frames

Figure 4.4.3 The outer wax and propolis layer of a brood cap

Figure 4.4.4 The inner silk layer of a brood cap

This silk forms the building material used to create bee tiny incubator cells to sustain and support the growth and development from pupa to an adult bee. This inner silk skin is hidden from view and rarely seen, unlike the silk made by spiders or caterpillars whose textiles are often on public display. Apparently bee silk is the most elastic of all insect silks. With practice it is possible to identify the wax and silk side of the cap without using a microscope. The silk side is more shiny and concave and the wax side is dull and convex.

Cocoon silk - golden threads

Find ten beekeeping books and use the index to search for cocoon silk. I tried this at home and was surprised to find that only one of these books thought cocoon worthy of inclusion in the index. Before I started this study, I hadn't thought too much about silk and only gave it the briefest of mentions while discussing the stages of bee development to students, or the damage caused to cocoons by wax moths. I regret not giving more time to discussing the structure, function, manufacture, and properties of this incredible material. Honey bee silk deserves a mention or even its own book. This fine elastic silk provides an engineered solution to support metamorphosis from larva to adult bee. It's difficult to think of a process more important in the development of a bee. So why do so few people talk about cocoons when they talk about bees? The reason is obvious, the silk made by honey bee larvae is hidden from view by a wax cap covering a brood cell. In contrast, most people on planet earth will have marvelled at a spider's silk web.

The first time that I had seen the golden silk threads synthesised by honey bee larvae was while reading Bees and Beekeeping written in the nineteenth century (Cheshire, 1886). The author used microscopy to create 70 detailed anatomical drawings with descriptions of the anatomy and physiology of honey bees. I was pleased to find four references to cocoon silk in the index, along with a picture (see figure 4.4.5) and description of how larvae spin their cocoons. These considered hand drawn images and descriptions are a testament to the value of detailed observation using microscopy – a skill which seems to have been lost in an age of digital imagery. How many people would choose to spend a whole day drawing an image when a phone camera can produce a higher quality image in seconds. Looking at hand drawn anatomical drawings I cannot help but feel that modern camera technology has significantly reduced the opportunities for this kind of mindful observation. The author also described a method to observe the 'golden' silk threads. The method was simply to dissolve the wax from a brood cap using an appropriate solvent and use dark field microscopy to reveal cocoon threads. Figure 4.4.6 shows my first efforts. Looking down the eyepiece of the microscope, I was immediately struck by this marvel of bioengineering and realised that I didn't know enough about the production of this fine textile.

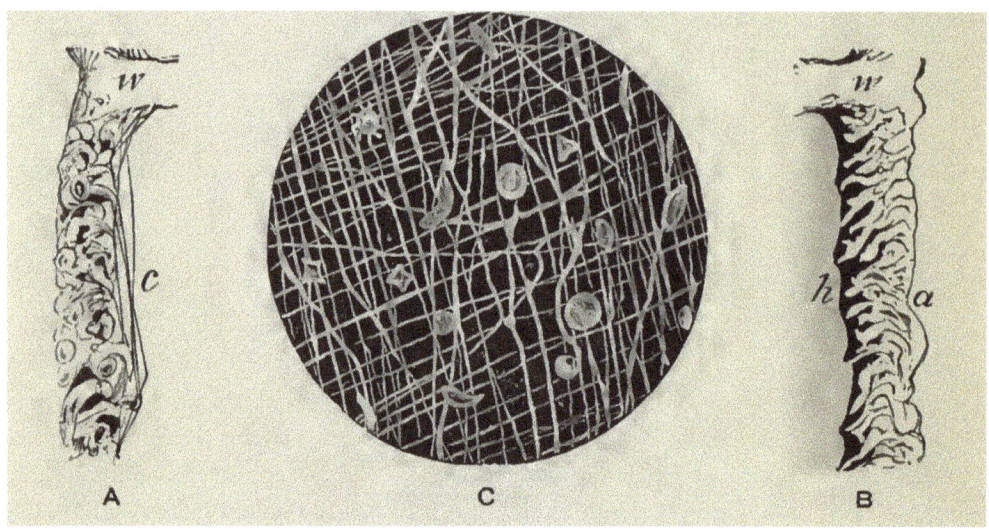

Figure 4.4.5 Cell cappings by Frank Cheshire

A sealing of brood cell, B sealing of honey cell, C sealing of brood after the wax had been dissolved, W wall of cell, H honey

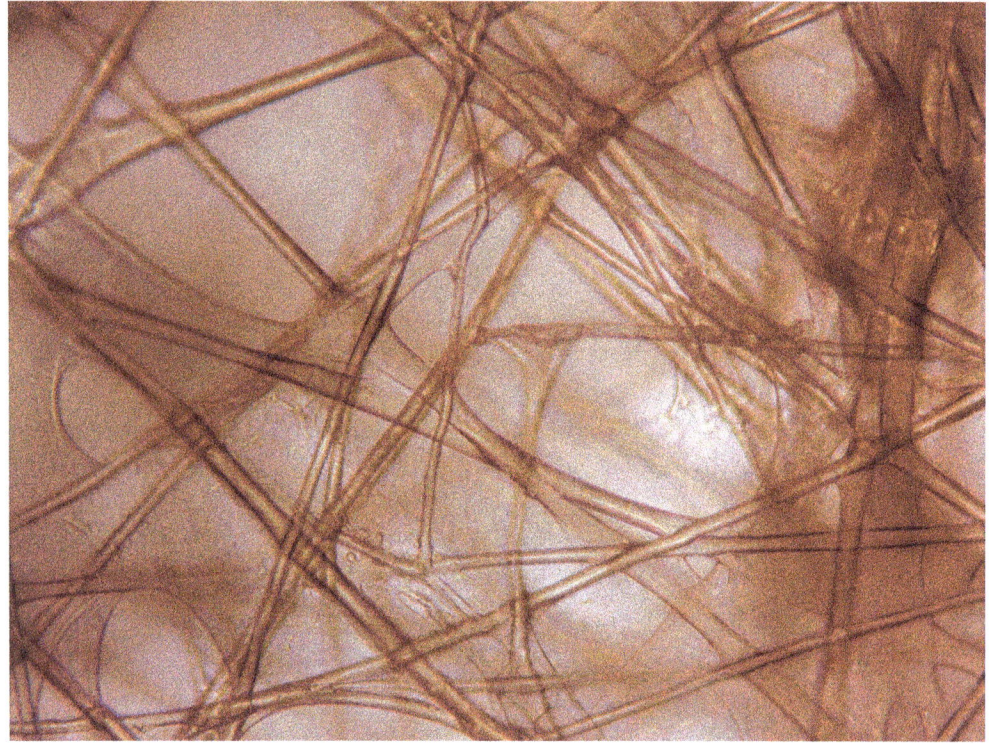

Figure 4.4.6 Cocoon silk fibres under microscope

This extraordinary material is produced by different insects. Silks of different kinds are used in reproduction, shelter, hunting, and, with bees in mind, a structure to support the remarkable process of metamorphosis. Spider silk is incredibly strong and can withstand the enormous impact of flying insects that get caught in their web and the huge forces spent trying to escape. Honey bee silk is the most elastic of all insect silks and can be extended by more than 200% (Poole, et al., 2013). It is the ideal material for honey bee larvae about to undergo complete transformation into a bee.

It is these mechanical properties of silks that give them economic value for fashion, medical supplies and more. Honey bee silk is no exception and has been successfully synthesised by laboratories using cloned honey bee genes responsible for silk manufacture. While this is undoubtedly a remarkable feat of human ingenuity, more incredible is the knowledge that many of the genes responsible for honey bee silk production have remained unchanged from far back in evolutionary time. The idea that the genes have been conserved for more than 150 million years in all investigated bee, ant and hornet species is astounding (Sutherland, et al., 2011). To think that these genetic sequences have been copied billions of times and reliably translated into silk and only expressed for two days in the lifecycle of a bee, is mind boggling.

Six honey bee silk genes have been discovered to code for the four proteins that form honey bee silk (Sutherland, et al., 2006). Just before the brood cell is capped, the six larval genes of the larvae are turned on. These glandular cells produce a fine colourless silk about 3 microns in diameter. The silk is manipulated by mouthparts and within a few days, the cell walls are covered by thin sheets of silk. Once the cell is capped and silk production has ended, these silk genes are turned off and new genes are expressed resulting in the silk glands becoming salivary glands. The silk is believed to provide a longitudinal cue for larval orientation, so that the head is nearest the capping. By the end of the spinning, the cell walls and inside of each cap are covered by a thin sheet of elastic silk. A comparatively small amount of silk in the debris suggests that much of this silk may be consumed by the emerging or adult bees.

Chalkbrood concerns.

Looking at the inspection board (figure 4.4.1) it is possible to see an increasing number of white chalkbrood bodies and a 'blackening' of the debris caused by its reproductive structures. Figure 4.4.7 shows that it is easy it is to spot the black spore-producing fungal structures and identify the areas of the colony most likely affected. It is conceivable that the presence of these black infectious spore structures is a result of bee hygienic behaviour. Less clear is whether these structures are left over from the previous season, or a consequence of a more recent infection. It is interesting to note that The National Bee Unit UK (NBU) says that typical symptoms start to appear in early spring as the colony starts to build up its population. Also, conditions such as damp and cold weather promote fungal spores. Certainly, April has been a very wet month, with lots of local flooding. With this in mind, and concerns about the health of the colony, I removed the mouse guard at the entrance, in the hope that it would give a little more ventilation to reduce the internal humidity of the colony. The NBU says that while chalkbrood is fatal to infected larvae and can result in a decline in bee numbers and honey production, that it does not usually cause the demise of the colony (Ponting & Stainton, 2020).

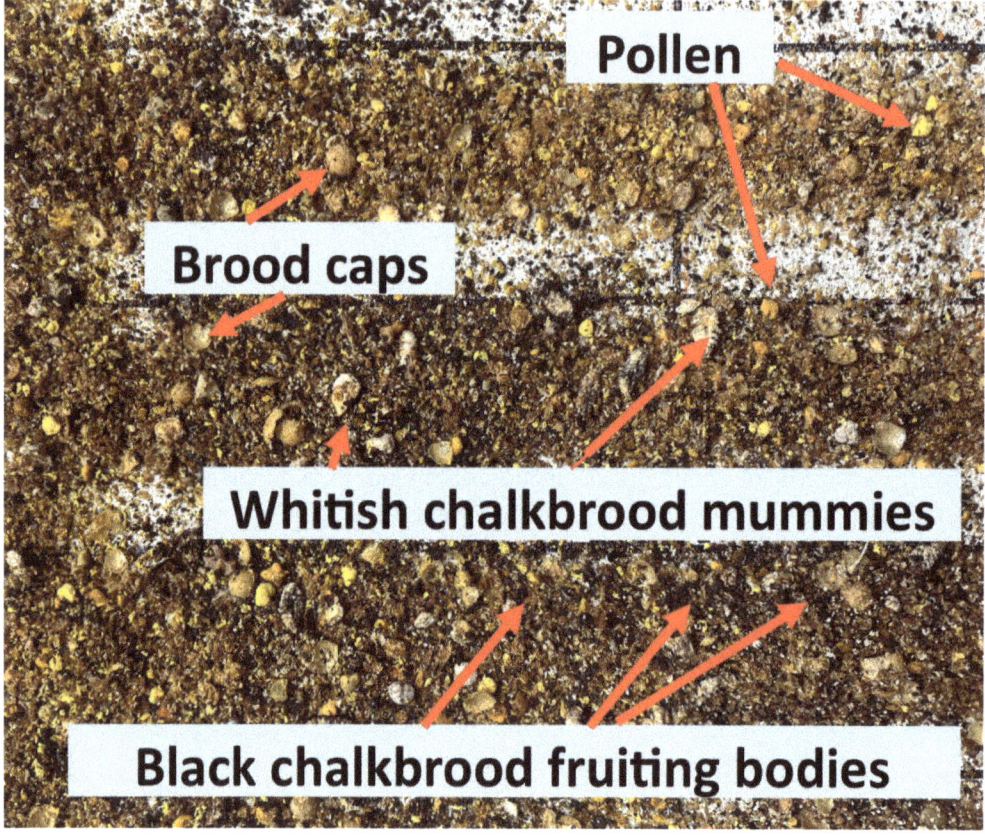

Figure 4.4.7 Chalkbrood in the debris

Pollen pellets in the debris

The sampled debris contains many 'freshly' foraged pollen pellets. It's a good sign because foragers primarily depend on the number of larvae present in the hive and the amount of stored food to make the decision about the rate of pollen foraging (Ghosh, et al., 2020), which implies that the more pollen found in the debris the more brood there will be. Nurse bees eat lots of pollen to synthesise royal jelly from their hypopharyngeal glands. The jelly is fed to young larvae for the first few days of development. Later, the larval diet becomes a mixture of whole pollen, honey, and enzymes. The larvae continue to eat this mixture until they spin their cocoons. The queen consumes royal jelly throughout her development. Figure 4.4.8 shows some of the new pollen found in the debris in April.

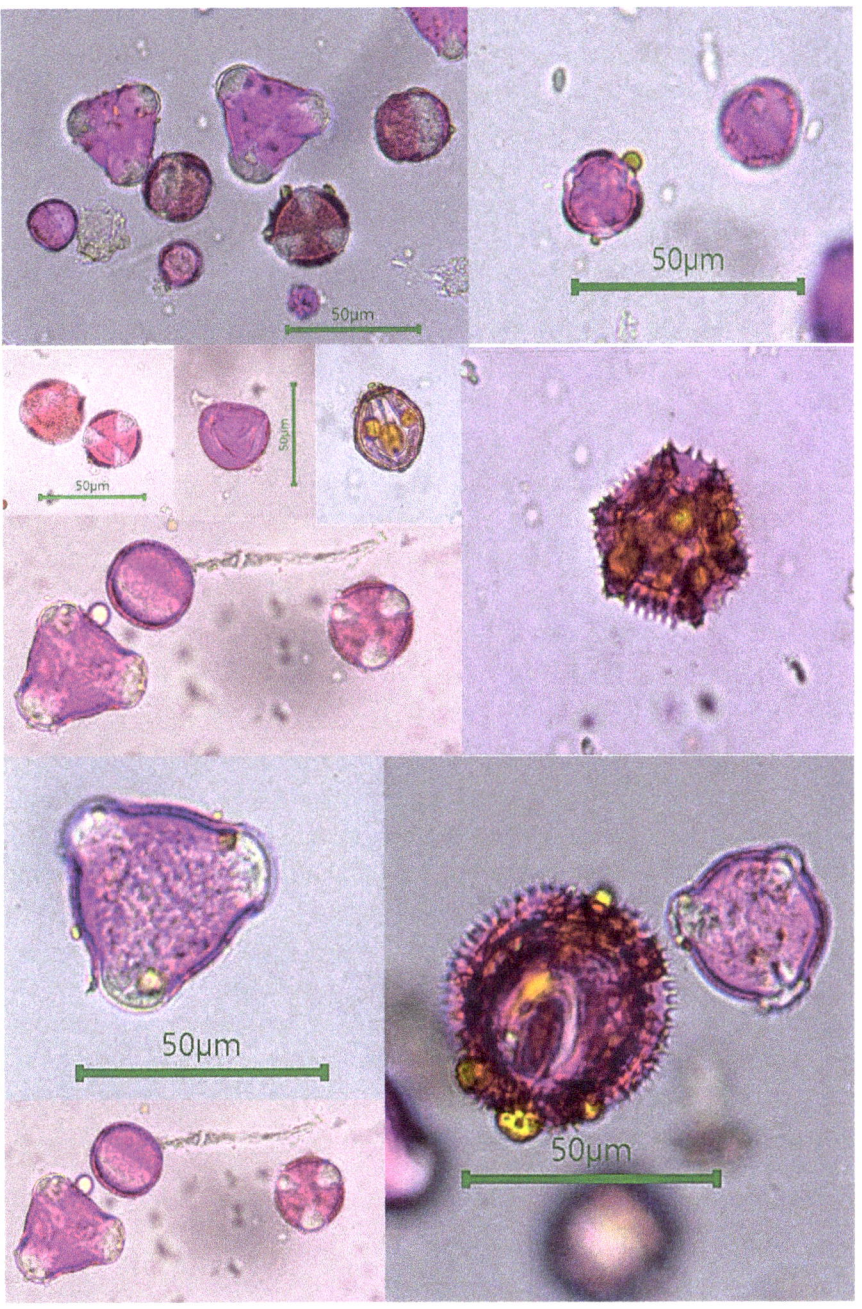

Figure 4.4.8 April pollen

Top left shows a mixture of rapeseed, plum (triangle shape), willow (top right), rapeseed and daffodil pollen (below left). Middle far right is the hexagonal structure of dandelion pollen. Bottom right looks like an unusual view of Asteraceae, perhaps a thistle, daisy or sunflower.

The many pollen pellets found on the inspection board help to remind us that debris is not the same as waste. A human intervention in the form of a metal ventilation mesh wastes both bee time and energy. A mesh trap that not only restricts access to pollen, but also for recycling and/or the removal of potentially harmful pathogens from the hive is unhelpful. It seems plausible to say that there is an upward risk of contamination from the inspection board to the colony. In the case of chalkbrood, the spores are very resistant indeed and may persist for at least fifteen years. This persistence means that regular cleaning of inspection boards and all beekeeping equipment is best practice to reduce infection risks.

In April, inside the bee shed had a pleasant aromatic smell from recently foraged propolis. Propolis consists of plant resins collected by foragers who visit the buds and leaves of resin producing plants, some known species include pine, birch, ivy, poplar, alder, oak, and chestnut. Foraging bees use their mandibles to break open plant structures to access these resins. Figure 4.4.9 shows a stellate trichome gland cell which is a common find in April. I wonder if this structure is responsible for resin production, but I have been unable to find anything in the literature to support this. Many plant species have their own associated trichomes. It is interesting to speculate about the potential to use trichomes to identify propolis sources in a similar way to how pollen identification is used to understand the pollen diet collected by foragers. I last saw a trichome in the debris in October 2023 when they were a common find.

In the hive the propolis is either stored, used raw, or mixed with wax for a variety of different purposes. Understanding the role of these resins is an emerging and important field of bee science. It is possible to see and smell tiny bits of raw propolis in the waste. Also, as in previous months, it is possible to find many processed pellets which are an amalgam of recycled wax and propolis. These bee pellets are often found entombing parasites (figure 4.4.10) and other objects that have the potential to decay and become a source of infection. A kind of hygienic behaviour that works to reduce the number of pathogens and promote colony health. Figure 4.4.11 shows a wax/propolis pellet encasing the main structural vein of the wing called the costa. Figure 4.4.12 shows an entombed fragment of exoskeleton. It is unclear why some of the debris is sealed inside these bee pellets and others are not.

Figure 4.4.9 A stellate trichome

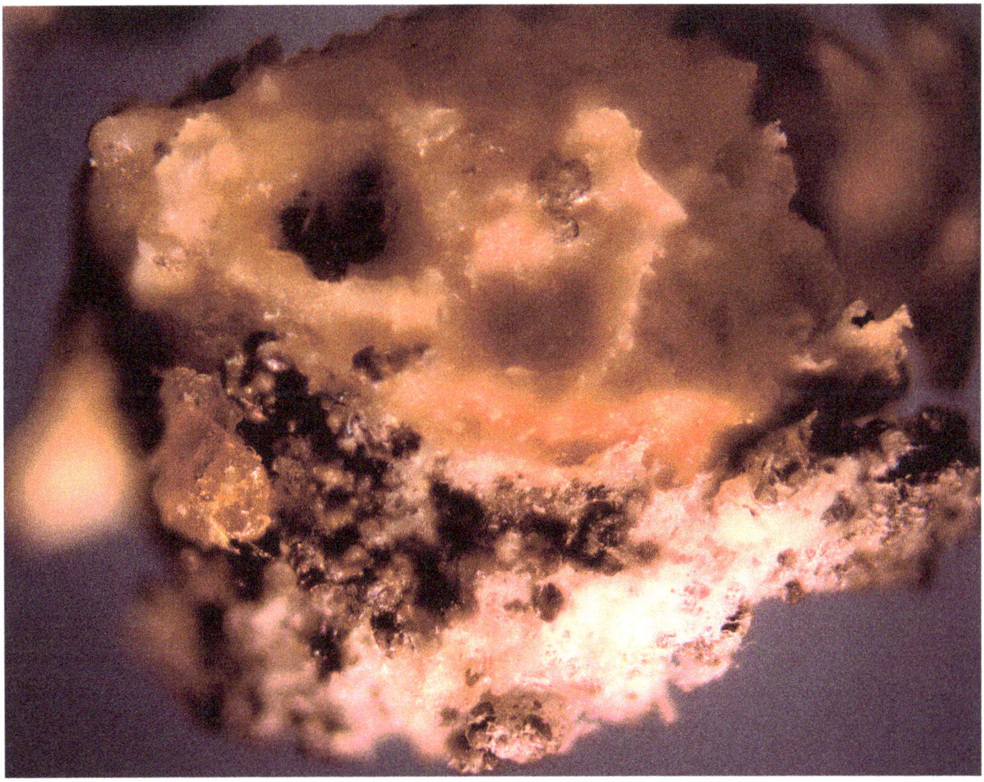

Figure 4.4.10 Propolis entombing fungal cysts

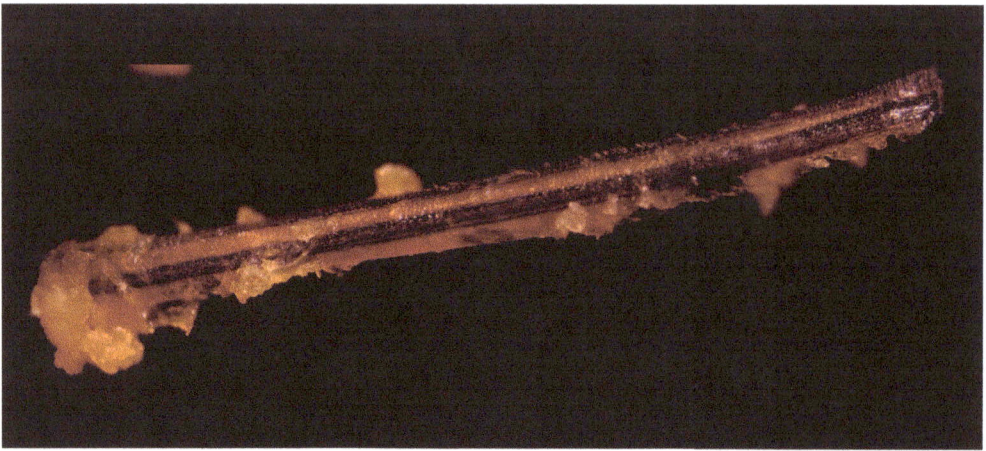

Figure 4.4.11 Propolis entombing a costa

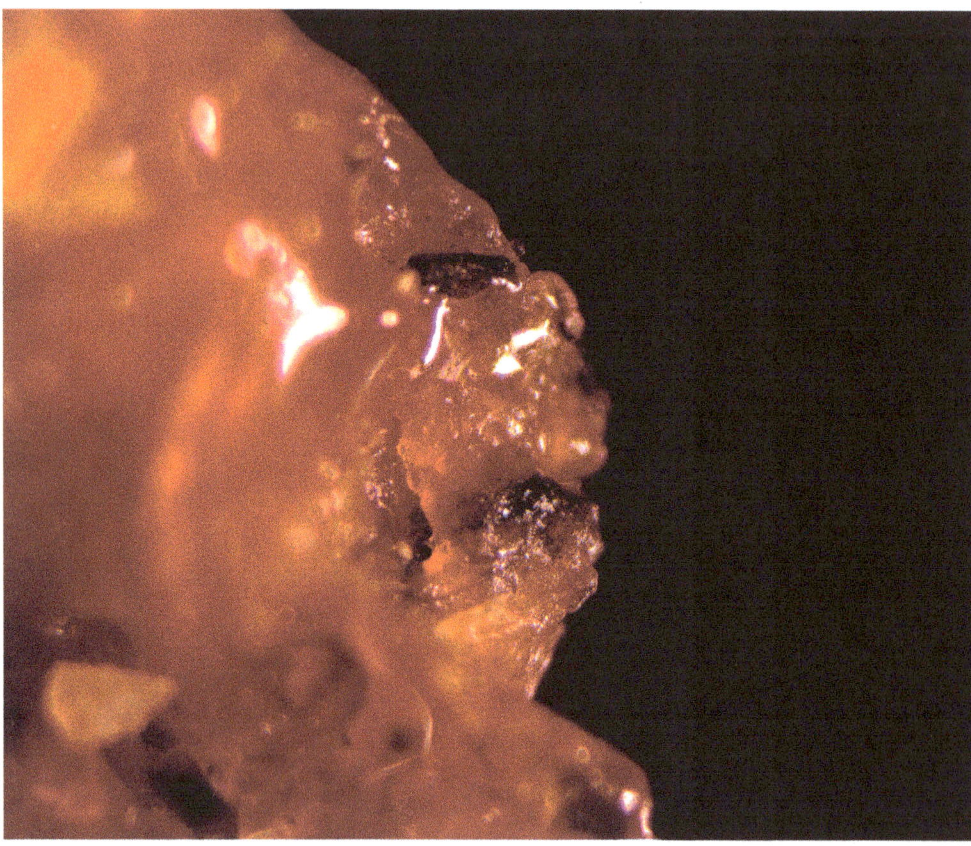

Figure 4.4.12 Propolis entombing bee exoskeleton fragments

4.5 May

Moths and Microplastics

Honey bee debris collected between 1st and 28th May

Minimum temperature 4°C

Maximum temperature 23°C

Mean temperature 13°C

Figure 4.5.1 shows a more even distribution of debris and colony activity in May. This is more than a 1300% increase in debris compared with January (figure 3.1). The debris pattern and number of intact brood caps shows that hive activity is now evenly occupying most of the lower brood box. I am surprised by the number of brood caps that manage to fall between the tiny holes of the inspection screen. I hadn't expected to find so many and imagine that, if left these would become an ideal habitat for the debris mites (and many more organisms) as discussed in previous chapters. Finding so many makes me reflect on how much the size of the ventilation mesh will be a determining factor in the size of the debris particles that get trapped on the inspection board.

The sheer quantity and equally spaced distribution of brood caps in the debris suggests that many of the cells will now be vacant for preparation before the queen starts laying in these again. I am now pleased that in April I gave this colony extra space with an additional brood box for expansion. I had considered splitting the colony, but for the purpose of this project, I was curious to observe the debris from a 'natural' cycle of beekeeping.

Hundreds of brood caps have fallen through the mesh of the ventilation screen, I imagine thousands of empty cells and worker bees preparing, cleaning and polishing the inside of cells to provide an ideal environment ready for the queen to lay a fresh egg in each cell. In the summer, these brood cells may be polished and reused every 21 days. Shortly after a queen lays an egg in a brood cell, the uncapped larvae will be visited and fed hundreds of times each day. This daily footprint, propolis and old cocoons, results in brood cells developing a blackened polished appearance. I wanted to find out if these brood cells were being reused, so removed a few frames directly above the seams producing the most brood caps (figure 4.5.2). I was delighted during this inspection to see an even pattern of polished cells with eggs and young larvae swimming in pools of royal jelly. Looking at those dark empty cells, each one a tiny incubator that supports the life of a keystone species, I am awestruck.

Figure 4.5.1 May debris

Figure 4.5.2 eggs, young larvae and polished cells

The value of blackened cells is sometimes contested in beekeeping circles. For some these black polished brood cells are a highly valued resource, while others are more vigilant in replacing them as part of their hive hygiene strategy. While queen bees seem to happily lay eggs in these polished cells, and tradition says they are an effective lure to attract swarms in bait hives, the size of each cell reduces slightly with each brood cycle. The wax also absorbs harmful pathogens and pesticides which may be brought into the colony via pollen and nectar from flowers. A study carried out in 2021 concluded that the residues of twenty-two pesticides and veterinary drugs were detected wax produced in Europe (Wilmart, et al., 2021). The most pressing research questions lie in determining the colony exposure and to better understand the pervasive exposure of these compounds (USDA, 2012). It seems logical to assume that pesticides will accumulate over time and perhaps evidence of this can be found in bee debris. At some unknown point it seems inevitable that these tiny wax incubators will become uninhabitable and unsuitable for healthy brood development. It seems likely that bees will have a very good sense for 'the range of this habitable goldilocks zone' and will have feedback mechanisms when a sub optimal point has been reached. The goldilocks principle has been adopted from the children's story "Goldilocks and the Three Bears", in which a young girl called Goldilocks tastes three different bowls of porridge and finds she prefers porridge that is not too hot or too cold, but just the right temperature. The analogy of "just the right amount" is frequently applied in the biological sciences where living things often have a 'sweet spot' where they can thrive. The Goldilocks analogy is a poor example when talking about pesticides because it would seem highly unlikely that there will be any pesticide value that is beneficial for honey bees.

Figure 4.5.3 Different-coloured debris

Figure 4.5.3. shows a debris mixture heaped neatly into a pile. If it were a spice mix, it looks like it could add a zesty tang to any dish. The different colours of the debris partly show different brood cycles. Fresh white wax scales are often found directly below cells that are being engineered, with each progressive brood cycle producing darker debris. The light-yellow debris was deposited beneath a fresh frame of foundation added to the brood chamber. This debris was collected in a different month but helps to illustrate how colour may indicate a different number of brood cycles.

The NBU (UK) advises that it is best practice to replace brood comb at least every three years and more frequent exchanges are advised if you have had disease in the colony or used varroacides. A three-year theoretical maximum of over 50 brood cycles in three years sounds like a lot (constant use is not possible due to many variables that will affect the rate at which eggs are laid). Assuming a more realistic number of brood cycles in three years is closer to half this value, I would love to ask the bees which side of the habitable goldilocks zone 25 brood cycles would be. I think the bees would take the advice of the NBU.

At my home apiary, I am working towards a rolling cycle of frame replacement. This involves replacing one third of brood frames each year. Removing the frames that have had the most brood cycles, the ones with the darkest and smallest cells. In my Slovenian hives, with double brood boxes, this involves removing 6 frames each year and replacing these with starter strips or new foundation. I am intrigued, about the possibility of using the debris to quickly identify the frames that have been used the most for brood rearing. I am also keen to develop this strategy to preserve the propolis that has been laid down from the previous season.

Research has demonstrated that propolis resin is spread on the inner walls of each brood cell (Pusceddu, et al., 2021). A team of scientists carried out chemical analysis of combs that had been polished ready for egg laying (oviposition) and of the 30 or so compounds identified, Kaempferol was the most abundant compound. In the plant world this compound is used to regulate growth and for defence purposes. In the hive these compounds are well known to have a direct effect on several hive pathogens. The dark propolis may look dirty, but it plays an essential role in reducing the microbial load of the colony by reducing the fungi, bacteria, and viruses present in brood cells. It is interesting to note that in clinical trials the most abundant compound found in propolis has exhibited antibacterial, antifungal, antiprotozoal, anticarcinogenic and anti-inflammatory effects. Many of the plant derived compounds found in propolis are today recognised as having novel potential for drug design (Periferakis et al. 2022). Unfortunately, smooth brood boxes do little to stimulate propolis collection and many beekeepers work to actively remove it from the colony. As a result, many colonies are propolis poor. Research suggests that propolis-rich environments contribute to hive homeostasis and support the practical implementation of rough box hives to support honey bee social immunity (Shanahan, et al., 2024).

It is generally believed that a combination of debris from footfall, pollen, propolis, pollen and cocoons make brood cells darker over time. I took some comfort knowing that the antifungal properties of the propolis coating the interior walls of brood cells may be reducing the fungal load in each cell. Especially considering the large increase in the dry black spore structures found on the inspection board in this month, which had darkened the overall colour of the debris. I was starting to worry about the health of this colony and its ability to deal with such a high level of infection. Figure 3.4 estimates that there were around 2.5 million of these structures in the colony. Figure 4.5.4 shows just one of these cysts. At 400 x magnification these structures resemble 'bags of footballs', with each one containing many spores. These spores are visible at 400 x magnification. Figure 4.5.5 shows an example of these black fungal structures on the inspection board.

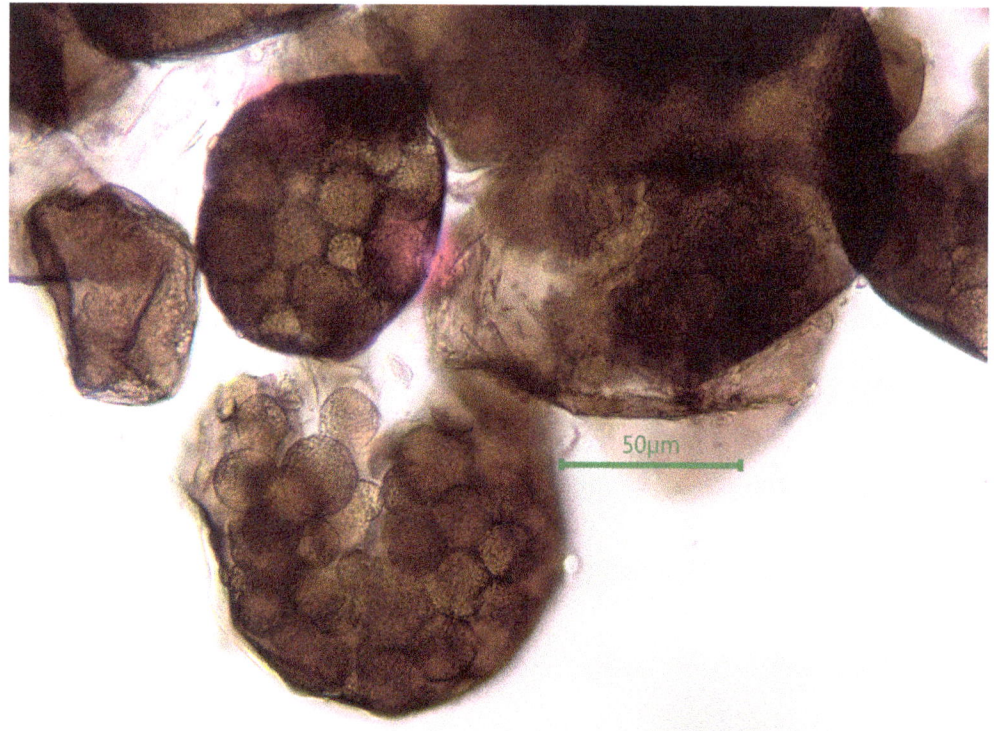

Figure 4.5.4 Individual chalkbrood cysts

Figure 4.5.5 Black fungal structures on the inspection board

While I am concerned about the effects that this stress factor may be having on the health of the colony, I'm choosing to believe that the bees are managing what appears to be a very high fungal load. I started to read more about chalkbrood but was unsuccessful in finding any published work stating a lethal dose of infection. In April, steps were taken to improve hive ventilation, but as the colony expanded so did the number of fungal spores. Monitoring chalkbrood in debris is new to me so I am unsure how much of a problem this is. BeeAware state

"Chalkbrood is not usually a serious disease among strong healthy colonies. However, in smaller colonies or those under stress (for example suffering heavy Varroa infestations) it can become a problem. The best method for keeping chalkbrood to a minimum is the maintenance of good strong stocks which appear better able to resist the fungus. Those colonies which are susceptible can be re-queened. Avoiding damp apiary sites will also help to minimise the effect of chalkbrood in colonies" (BeeAware, 2024).

So, for now, I am reassured that the colony appears strong, with a small varroa population and that the bees seem to be doing a great job at removing the spore structures. Perhaps the presence of so many spores in debris is a good indicator of hive hygiene in practice.

One of the joys of writing this book has been a monthly meeting with the editors Ann Chilcott and Christine Coulsting. We have developed a routine of discussing some of the find's month by month. In May, chalkbrood became the main topic of conversation. One idea discussed was that as a colony begins to expand early in the season, that there isn't an adequate nursing population to manage the rapidly growing area of brood and that this can create difficulties with homeostatic regulation of temperature and humidity, which can in turn create ideal conditions for fungal growth and reproduction. Add to this scenario a recent period of damp and cold weather (which we have had) and the problem becomes worse. Additionally, individual queens may be genetically disposed to thermoregulate the colony at different temperature and humidity values. The logical conclusion to this argument would be to requeen the colony. Both are interesting ideas, but for now I will leave the colony alone and continue to monitor.

Pollen

Figure 4.5.6 shows the pollen found in the debris. It is satisfying to know that, apart from the field beans, we have planted these pollen producing flowers on our flower farm. It's not surprising to find that planting flowers is good for bees. Starting top left and row by row the pollens identified in the debris include; what looks like dead nettle (difficult to confirm due to a lack of visible pores), poppy (top middle), field beans, broom, honeysuckle, apple blossom and aquilegia.

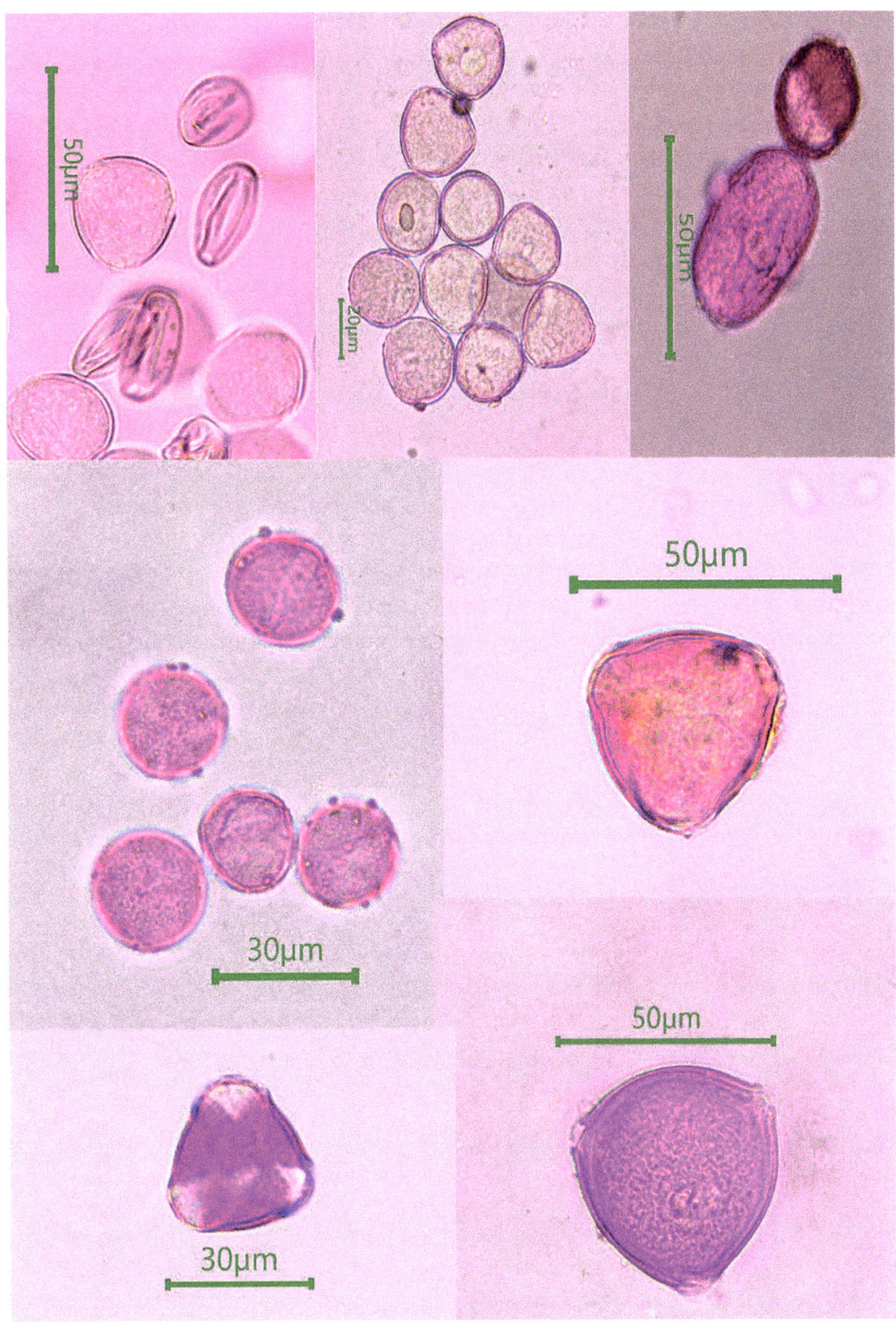

Figure 4.5.6 May pollen

We have been planting flowering plants on our smallholding for almost 15 years and the results are starting to show in many ways. The pollen found in the debris reflects some of those changes. Our overall goal has been to make a living from growing and selling cut flowers and to provide enough forage for a few colonies to survive without relying on supplemental feeding, as well as creating a beautiful place to live. To plant flowers instead of feeding sugar was my plan. It probably makes no financial sense at all, but the principles of sustainability and ensuring a balanced diet for pollinators seems like a very good aspiration indeed. However, it takes research at an individual apiary level to better understand the seasonality of local forage. Pollen analysis of debris can provide insights into the forage being utilised.

Planning for forage

Part of planning for a balanced diet of forage involves thinking about the needs of all pollinators, not just the bees. An interesting, but concerning idea, is that the domesticated honey bee contributes to a loss of wild pollinators through resource competition and spread of disease (Gonzales-Varo 2018). The conclusion is that the decline of wild pollinators, such as moths, bumble bees, and hover flies is partly because of a continual replenishment of honey bee stocks by beekeepers. It's an argument I have heard from young people at school who are often the most passionate wildlife advocates. It's a view that needs an open mind and to be carefully considered. If proven through monitoring, it may well necessitate controls on honey bee stocks.

The view of a flower farmer and moth advocate - honey bees on a flower farm

I have asked Paula Baxter to contribute to this topic, because she knows more about the foraging behaviour of pollinators around my apiary than anyone I know. Not from academic study, but rather the daily observations of a flower farmer for 14 years. She is up close and personal with flowers and pollinators on a daily basis. If I want to know what my bees are feeding on, I ask Paula. She is also passionate about other insects and especially moths, carrying out her own weekly moth trapping and reporting into the UK database. I asked Paula for her observations.

Paula's thoughts about my bees

Ray started keeping bees on our smallholding in 2010, not long after we moved here. At the time, the land was a mix of rough grassy scrub, perennial 'weeds', mixed young native saplings and water. The mill pond takes up over a quarter of our land and at that time it supported very few plants, with no submerged plants and a monoculture of sweet rush as a marginal. The farmland surrounding us was intensively cultivated for cereals for decades. The years since then have been a constant exercise in planting seeds, plants and trees.

The flower farm was established in 2012, with the aim of growing cut flowers for sale to florists and direct to the public. It's always been my philosophy to grow what grows well here, to plant varieties that thrive in our cold, wet soil. We've never used pesticides or herbicides and work in a way that has as low an impact on the environment as possible. We sell cut flowers and foliage across a long season, from March to the end of October, so it's important to be able to have as many plants as possible that flower across those months.

In some ways, the combination of keeping honey bees and growing cut flowers isn't as ideal as it might first appear. The flowers are generally (and ideally in business terms) cut and sold before they bloom fully, and pollinated flowers quite quickly degrade as they start to form seed. Many specialist cut flower varieties are double flowers, which are much more difficult for pollinators to access.

However, the sheer volume of flowers in a relatively small area, combined with a long flowering season would seem to benefit honey bees, along with many other insects. Working alongside them, it's notable that honey bees prefer different kinds of forage at different times. Although many plants are held to be bee-friendly, the flowers have to be at a fairly mature stage for the bees to be attracted to them. Once they are at the right stage the plant is covered and all that can be heard is buzzing. Other types of bees and pollinators seem to use a broader range of plants and over a longer timespan. Although Ray would maintain that we've planted a lot of things specifically for honey bees, the reality is that most of them are for cut flowers. There's very little that won't be cut if it's at the right stage and I have an order for that colour! But if it doesn't get cut and sold, a flower is quick to open and is fair game for all the pollinators that our plot supports.

It's been very interesting to find out through the pollen analysis which flowers are used the most by honey bees, but also the relatively limited number of varieties that appear in the pollen analysis. What is more startling is the increase in the diversity of other pollinators as the range of plants has expanded. I've been monitoring moths for the past 5 years and each year brings new species that feed and breed on particular plants. Last season we had one of the first Scottish records of mullein moth caterpillars on verbascum.

Figure 4.5.7 shows a few examples from more than 300 moth species found near my apiary.
Figure 4.5.7 Moths at the apiary

Starting top left and row by row - white plume, elephant hawk moth, mullein moth caterpillar, elephant hawk moth caterpillar, bulrush wainscot and small magpie.

A micro moth *Cnephasia longana* was recorded here, a first for our area. It feeds on wild carrot (*Daucus carota*) and sea thrift (*Armeria maritima* spp.), neither of which are naturally occurring in our immediate area, but both are grown in decent quantities as cut flowers. We sowed seed and now have thriving patches of reed mace (*Typha latifolia*) plus the bulrush wainscot and large wainscot moths that feed on it. If you plant it, they will come it seems. The idea that keeping honey bees will have an adverse impact on other pollinators might seem logical if pollinators all foraged the same resources at the same time, but, in reality, I haven't observed any evidence. The number and diversity of moths in particular increases year on year at this location, as the flowering shrubs and perennials mature to provide more nectar, pollen and habitat. Most of them feed on very different plants and at widely differing stages, often feeding as larvae in stalks and leaves, rather than purely on flowers. Although we have honey bee colonies, the numbers of bees as a proportion of the total that are supported here reduces each year. A casual walk around our pond on a sunny day will show thousands of micromoths, damselflies, butterflies, and invertebrates feeding on grasses, pollen, and each other. The answer to any concern over adverse impacts of honey bee colonies would seem to be to just grow more plants, and increase the diversity of plants. Though in my flower farming world, that could be the answer to most things!

Microplastics revisited

Back in January I started to find what looked like human made fibres in the debris. Initially I dismissed these, not quite believing that microplastics would find a way into my remote colony in the Scottish Borders. The discovery of a microscopic warp and weft textile (see February) in the debris entombed in propolis changed that view and I wanted to better understand the issue. I started to read everything I could find about microplastics and honey bees. I put a call out to beekeeping friends and previous students and asked if they would be interested in collaborating to understand the situation better, through a day of experimentation. A decision was made to run the event with young people in the first instance, and organise a later event for adults. It was just brilliant to be involved with such fantastic young people from three high schools working together to better understand the presence of microplastics, a reminder of the contribution young people can make to the debate and how much I miss working with the students at Bee Club.

I was concerned about making any conclusions using the results from just one hive. More replicates were needed to have any confidence about conclusions. The local high school was generous in offering us the free use of their laboratories, so it soon became a straightforward event to organise. We met on the 14th June 2024 at Berwickshire High School to carry out a DIY research project about this topic.

I also contacted Norman Carreck, a distinguished apiologist and lead bee researcher to ask if he was aware of any work being done in this area. Norman signposted me to the first study of this kind carried out in 2021, by a team of EU scientists. Norman is the UK coordinator for INSIGNIA-EU, which is an international collaborative project led by scientists from eleven organisations in ten countries, which aims to design and test protocols using honey bee colonies for the detection of pesticides, microplastics, heavy metals, and air pollutants. It is also interesting to note that their website (https://www.insignia-bee.eu/publications) produces guidelines for citizen scientists interested in using honey bee colonies for bio-monitoring of the environment.

The INSIGNIA-EU study (Edo, et al., 2021) sampled 18 Danish colonies and found 13 different types of plastic, categorised based upon their appearance into 4 groups of fibres. Using a similar method, Scottish students categorised fibres found in the debris into the same groups as used in the EU study. Our counting method was as described in section 2.3.

Figure 4.5.8 shows the students' analysis (Payne, et al., 2024 Supporting teachers - Angella Yekken, Lesley Rosher, Jen Addie & Ray Baxter). Like the EU study, students found fibres present in all the colonies sampled. The EU team found that fragments and fibres accounted for 90 % of the finds. In this school study, fragments and fibres represented 78 % of finds. Schools do not have access to expensive diagnostic testing as used in the EU study, but using the results of the Danish study it seems plausible that most of the Scottish finds will have been a mixture of polyester, polyethylene, polyvinyl chloride, and polyurethane. The youngsters absolutely loved being part of this day and it is pleasing that some are planning their own advanced higher biology projects to explore their own follow up questions.

If this small study looking at rural Scottish honey bee colonies identified microplastics in each sample, it seems very likely that every UK colony contains microplastics, an interesting hypothesis to try to disprove.

Textiles shed fibres at all stages of their life cycles. When they are worn, manufactured, disposed of and especially when they are washed. It has been estimated that a single wash load can release millions of fibres. With so many textile fragments released from a single wash, finding an estimated 45000 fibres in the colony in May starts to sound like a small number. Figure 4.5.9 shows an increase in the estimated number of fibres in my home apiary by month.

Since the first account of microplastics in atmospheric fallout reported in 2015, more than 70 studies have demonstrated the presence of microplastics directly in the atmosphere (Aeschlimann, et al., 2022). It is easy to imagine how small particles of fibres could result from erosion and by direct release from a variety of activities, such as washing machines. Also, how these particles could become further eroded and then transported by wind and rain. Average daily deposition rates have been estimated to be hundreds of tiny microplastics per square metre and these fibres can be transported thousands of miles before being deposited. No wonder honey bees become bio-samplers, collecting particulates, whether they like it or not, in the course of their foraging activity.

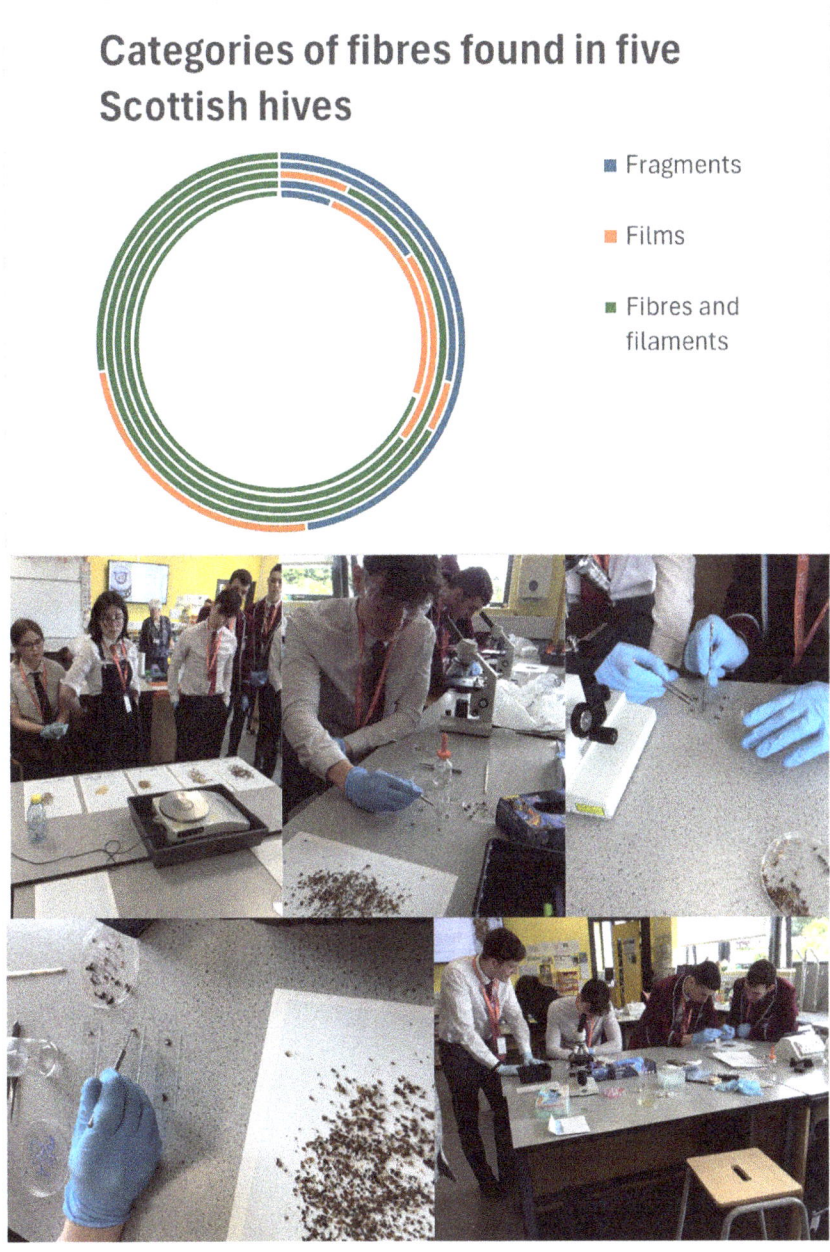

Figure 4.5.8 Categories of fibres found in five Scottish hives

Does it matter that microplastics are found inside the nests of an important keystone species like a honey bee? Certainly, having looked at lots of debris, it looks like honey bees are very good at collaborating to groom fine particles from their bodies. Perhaps they can deal with this grooming challenge, perhaps not. It's quite some unplanned experiment being carried out by exposing a keystone species to these fibres and waiting to find out.

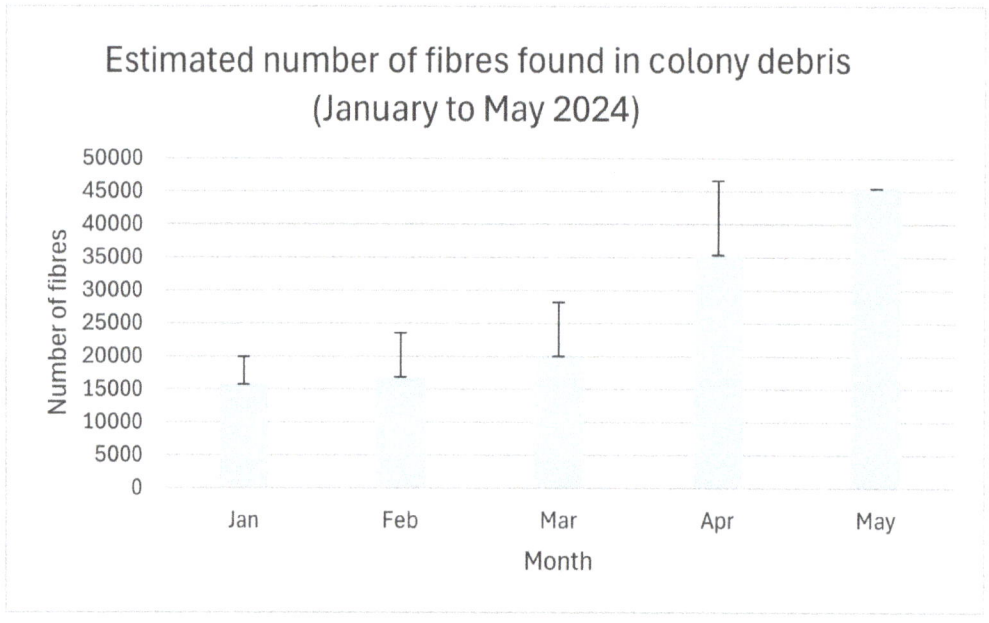

Figure 4.5.9 Fibres found in the debris each month

Interesting research about lead concentration within different parts of the colony shows a dilution effect happening, with most lead being found in adult bees (assumed to be from accumulative foraging), less is found in pollen and the lowest concentration is found in honey (Steen, 2016). I have wondered if something similar could be happening with the concentration of microplastics throughout the brood nest, but in this case, the hypothesis would be that most microplastics are found in the debris, followed by returning foragers and lastly in the honey. Research would be needed to find out if grooming causes a dilution effect, but if proven it would make a good case for using debris as part of the methodology for detecting environmental contaminants sampled by honey bees. Conversely, does grooming result in debris that represents a bioaccumulation of environmental contaminants? It is interesting to ask the question.

Research has demonstrated that bees transfer microplastics to different parts of the colony (Alma, et al., 2023). This study supplied a sugar solution with a known concentration of microplastic fibres in a sugar solution for one month. These fibres were then found incorporated into the digestive tract, cuticles, larvae, honey and wax. It is interesting that most of the fibres accumulated in the wax. A repeated observation throughout my debris study has been the presence of many fibres and hairs in processed wax. Could this support the idea that some kind of dilution/bioaccumulation effect may be occurring? Where grooming behaviour results in higher concentration of fibres in the debris compared with other parts of the hive. Also, if bees are acting as bio-samplers within a foraging area it would seem plausible that hive and debris will contain an accumulation of fibres from that area. While the topic of microplastics is concerning, fascinating and worthy of investigation, I have decided not to continue looking for these fibres. There really is enough to do in June!

4.6 June

The June gap

Honey bee debris collected between 1st and 28th June

Minimum temperature 3°C

Maximum temperature 24°C

Mean temperature 13°C

The inspection board (figure 4.6.1) shows many brood caps. Surprisingly there was a reduction in the total weight of debris deposited on the inspection board. This trend can be identified simply by comparing pictures of the two inspection boards. Figure 3.1 shows this to be a 28% reduction in debris - what seems like a significant change in activity. Figure 3.4 shows a 60% reduction in chalkbrood structures in the debris. The debris still looked very dark, but much less so than for May.

This is a change from previous months which have seen a continual increase in debris month by month. Something has changed. Is this the June gap? A time when the spring flowers are over and the summer ones are not in bloom. Beekeepers in the UK know that June is a risky time, because expanding colonies can starve due to a lack of nectar for the growing population.

The NBU website states that nibbled cappings can be a sign of starvation.

"Nibbled cappings with healthy looking pupae beneath are a typical sign of imminent starvation. Starving bees are said to nibble brood cappings in desperation, presumably in search of food. Also, that a starving colony will remove pupae and drop them on the floor or throw them out of the entrance" (National Bee Unit Website, Last accessed 05/03/25).

Figure 4.6.1 June debris

Clearly this is something to look out for in the debris as this behaviour would likely lead to an increase in nibbled cappings and discarded pupae in the debris. I couldn't find any evidence of these, only whole intact brood cappings (not nibbled) were visible. However, the reduction in debris weight was the only prompt needed for a quick inspection, just to check.

I once had a colony that starved, and finding dead bees with their heads in cells trying to feed was a depressing sight indeed. I resolved never to allow that to happen again. A quick look at the frames showed plenty of stores, so there is a different explanation for the reduction in weight of the debris in June. For now, I'm choosing to believe that this reduction in activity is somehow related to temperature (June has been a colder and wetter month than May) which somehow has caused a pause in egg laying and brood development. If this is the reason, then I would expect to find a reduction in the number of brood caps found in the debris. Something to look out for in July.

My apiary is in the middle of a flower farm, surrounded by hundreds of species of flowers in bloom, yet, only 4 kinds of pollen can be found in the debris during June (see figure 4.6.2). The lowest number of pollen species in this study so far. They are mostly field beans (bottom right 52%), rapeseed (left 36%), apple (bottom left 4%) and what may be alder (top right could also be a shrivelled pollen grain of something else). Field beans have been present in the debris from every month sampled so far. In previous months these pollen grains can only have come from last year's crop as the beans were not in flower. It is possible that they were contained in different areas of the hive, such as honey, pollen stores and bee bread.

Perhaps this reduction in the number of pollen species could be caused by the recent poor weather? However, there may be other explanations to explain this reduction of pollen diversity in the debris, such as

- Perhaps these four pollens are sufficient to meet the nutritional needs of the colony, so the bees have no need to go elsewhere. It seems logical that pollen quality trumps diversity. In terms of nutrition, the key thing is that the amino acids are present to build and repair tissue, produce hormones, create an immune system, and everything else needed for bee building and repair. It is interesting to note that research has demonstrated that field beans contain many of the essential amino acids required for bee health (Cote, et al., 2022). And there is a high percentage of field bean pollen in the debris. The same bean is often advocated as a human food because of its high protein content and well-balanced amino acid profile. Also, as mentioned earlier, the pollen from field beans is the only pollen to have been found in the debris for every month sampled, so it is clearly a favourite for pollen storage too. Figure 4.6.3 shows some of the flowering field beans near my apiary in June.
- Could it be that the reduction of pollen diversity in the debris is because the colony has a good amount of stored pollen from previous months, and in June

it prioritises the collection of nectar, so there are few bees collecting pollen for its own sake and high yielding nectar plants win out.

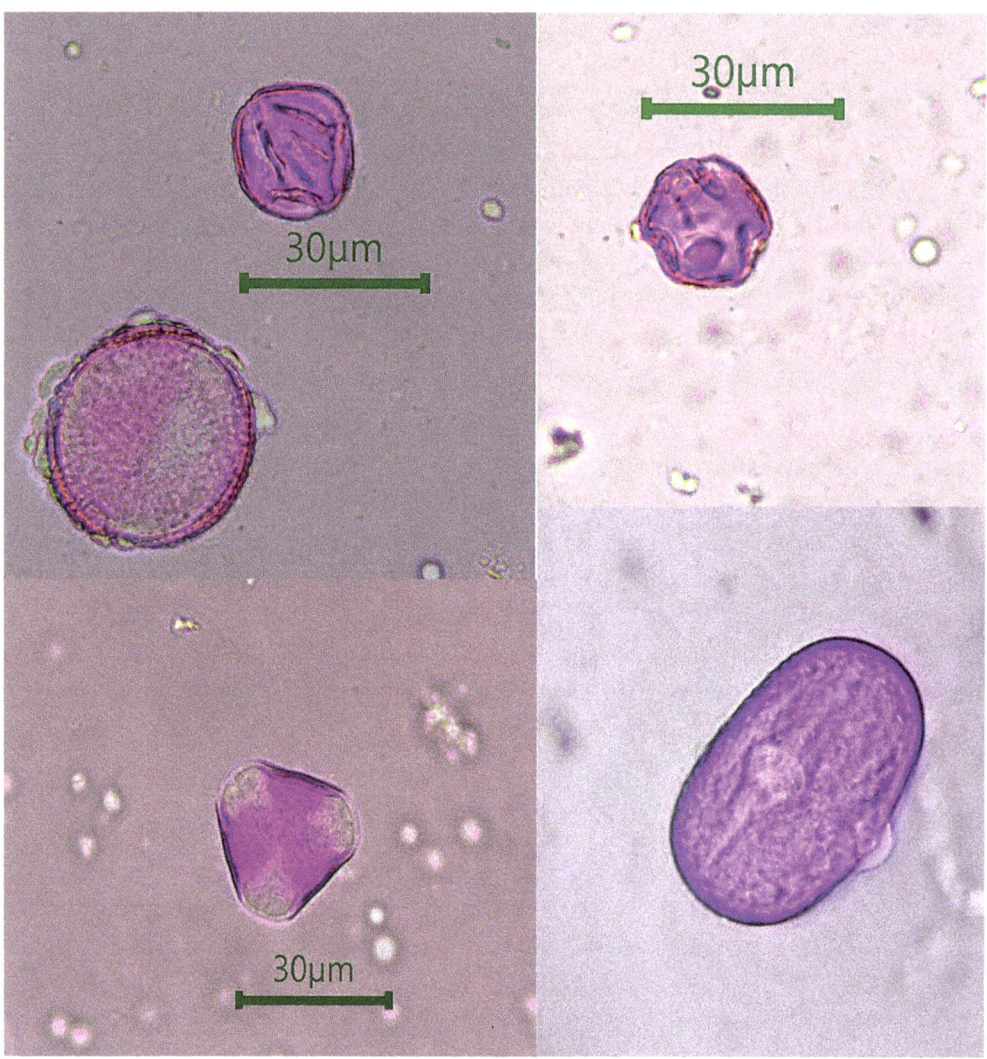

Figure 4.6.2 June Pollen

Figure 4.6.3 Field beans in flower.

- A different explanation could be suggested by research that shows that the way bees use pollen varies from month to month (Leponiemi, et al., 2023) and that pollen is distributed within the colony in different ways in different months. In my home study it is still unclear which process is producing the pollen found in the debris. Is it mostly from bee bread or from foraging? The fact that pollen is found in the debris every month, even when the bees have been unable to forage due to cold weather, shows that things are much more complicated than the assumption that the pollen found in the debris is from foraging. What proportion of pollen found in the debris comes directly from foraging, pollen stored and/or from bee bread is unclear. Although finding pollen lumps in the debris makes answering this question much easier as these can only have come from foraging.
- Finally, could the reduction in pollen diversity in the debris be an example of investment guided foraging in action, where 'time is honey'. Does this show bees actively seeking out and focusing on the crops with the biggest nectar/pollen return and using their energy to forage from those.

Swarm season

Imagine my joy at getting an early morning phone call.

"We seem to have an enormous swarm of bees in our front garden"

I think all beekeepers must be familiar with the saying

"A swarm in May is worth a load of hay; a swarm in June is worth a silver spoon; but a swarm in July is not worth a fly"

I had been hoping for a swarm colony to add to this project.

I made a quick visit and found what was actually a very small swarm. The home owner proceeded to say it was one of many that had come from the eves of the house. Even better I thought, a cast swarm, most likely with a virgin queen.

This gift of a swarm was worth much more than a silver spoon. My plan was to compare the debris patterns of bees from a 'virgin colony' in three development stages.

Stage 1 Pre mating - a colony with a virgin queen

Stage 2 Post mating and pre-emergence - A colony with eggs and larvae

Stage 3 Post emergence - a colony with capped cells after 21 days

Before lunch, the colony was rehomed and fed a sugar syrup solution. Figure 4.6.4 shows the swarm after a day quickly drawing out new comb from a starter strip, and figure 4.6.5, shows this new frame being added to hive above an inspection board.

Figure 4.6.4 Drawing comb

Figure 4.6.5 Fresh comb above the inspection screen

A summer wouldn't be one without catching a swarm. Beekeepers are trained to discourage and control swarming. It is argued that it is in the best interests of the bees, the beekeeper and the public to prevent swarming. The reasons given are that fewer wild swarms survive than managed colonies, the beekeeper will lose many bees, and in turn have a lower honey yield, and also the general public are often frightened by swarms. It is a challenge for the beekeeper trying to work with a colony if half of it flies away.

However, swarming is essential for the natural reproduction of the colony. It is a remarkable biological phenomenon where the colony can split into many sub-units with the potential to become new colonies. The first swarm is usually headed by the old egg laying queen and is called the prime swarm. The following swarms are headed by virgin queens. These casts are often regarded as bad news for the parent colony as they can rapidly deplete the population.

4.7 July

Swarm debris included

Honey bee debris collected between 1st and 28th July

Minimum temperature 6°C

Maximum temperature 23°C

Mean temperature 15°C

Figure 4.7.1 shows the debris pattern on the inspection board for four consecutive weeks in July. In the third week of July no brood caps could be found. This change, accompanied by an increase in debris weight, suggests a major event in the development of the colony. It seems fair to conclude that a total lack of brood caps in the third week of July shows that there was no longer a laying queen. My best guess was that the poor spell of weather as reported in June led to a pause in the queens laying schedule. Keen to understand the situation better, I lit a smoker and carried out the briefest of inspections. Using the debris pattern as a guide, I focused on removing frames directly above the deepest/darkest debris of the colony. I was reassured to find lots of open larvae and eggs, but no capped brood. Enough evidence to show that the colony did have a laying queen. I was also reassured by the number of bees and how they were filling a honey super, which showed that they had not swarmed. However, I couldn't rule out a supersedure event as a possibility.

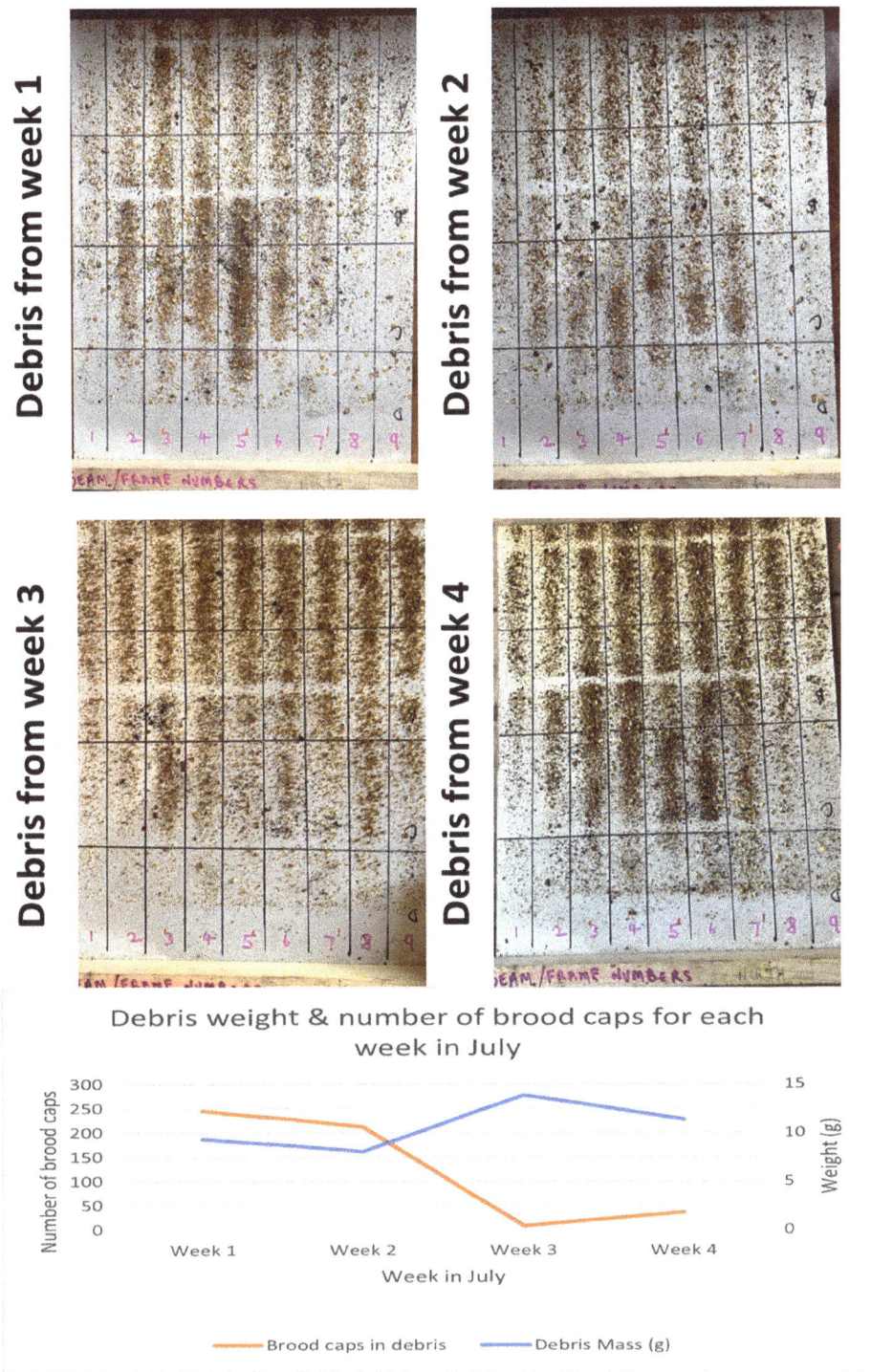

Figure 4.7.1 July debris - sampled weekly [whole page]

Most pleasing was that watching the debris helped me to spot that this event had happened. A change to a weekly sampling strategy in the summer months was necessary to keep track of changes, due to the sheer weight of debris. Repeat photopoint monitoring (in this case taking photographs every seven days) has proved very useful. It is not a new concept as 'repeat' photography is the foundation of many citizen science monitoring programmes. For example, in one UK study citizen scientists contributed images of the coastline to detect and map changes to the coastline. Clearly, this technique has only become possible due to easy access to digital devices that can take high quality photographs. A technological trend that will continue to develop as more sensing and diagnostic capabilities are created, thus providing more opportunities to study biological systems such as honey bee colonies.

This is well illustrated by figure 4.7.2 which shows a comparison of the debris from week 2 and 3 in July. One week later there were lots of whole brood caps and in the following week there were none.

Figure 4.7.2 Debris change in one week - recording a major event

Figure 3.1 shows that there was a 15% reduction in debris deposition in July. As a proxy indicator of bee activity I wonder if this reduction was because of the event shown by figure 4.7.2. The heat map (figure 3.2) also suggests that this event had a major effect on colony development and suggests a decrease in brood compared to the previous two months.

In the same week there was also a reduction in the number of chalkbrood structures as shown (see figure 3.4). Could this simply be because there had been a break in the brood cycle meaning there are fewer opportunities for the chalkbrood fungi? If so, could this brood break be beneficial for the colony?

Thinking back to 21 days earlier, the colony was experiencing a period of dreadful weather. It is known that the rate of egg laying can vary depending on several factors. The natural rhythm of the colony cycle, a mix of internal variables and the external environment can all help to explain why a queen bee stops laying eggs in midsummer. The availability of resources and the size of the colony are important considerations, as the queen regulates her egg laying to maintain optimal conditions. There are of course other reasons that may explain a reduction in egg laying. I am choosing to believe that the most likely explanation is the spell of dreadful weather, with consequential effects on forage availability being the most likely causes for this change in queen behaviour to attempt to regulate an optimum population for the environment. Whatever the explanation, it was pleasing to see that by week 4 brood caps were again found in the debris, albeit at much lower levels than previously recorded.

Propolis

It has been a welcome discovery that the ventilation screens on the rear of the AZ hive naturally seems to induce high levels of propolis deposition. I have read that a propolis envelope in the hive can play an important role in bee health reducing pests, parasites and microbial pathogens. As discussed earlier, numerous papers have shown that propolis contributes favourably to honey bee immune response and colony social immunity (Hodges, et al., 2019). The more I read about this substance, the more convinced I am about the importance of propolis for bee health. I was keen to support my bees to help them to build a propolis envelope and had read that it was simple to increase the rate of deposition by making rough surfaces on the inner walls and increasing the textural stimuli available to bees - it was on the to-do list! To discover that the AZ hive seems to do this by design saved a job. These ventilation screens are designed to be easily removed and would make propolis harvesting quite simple, so perhaps they were designed with this in mind.

Figure 4.7.3 and figure 4.7.4 show changes in the propolis being stored in July. These pictures show that propolis has been stored throughout June and July, with the first propolis always deposited near the edges of spaces on the ventilation screen and gradually working inwards. Also, that there are many different shades and textures, making me wonder if each colour tone is an expression of the local floral signature, a reflection of the plant resins creating a unique identity for each hive. Certainly, it would be interesting to identify the floral contributors to this propolis. Like honey, each type of propolis will have a unique blend of flavours and properties.

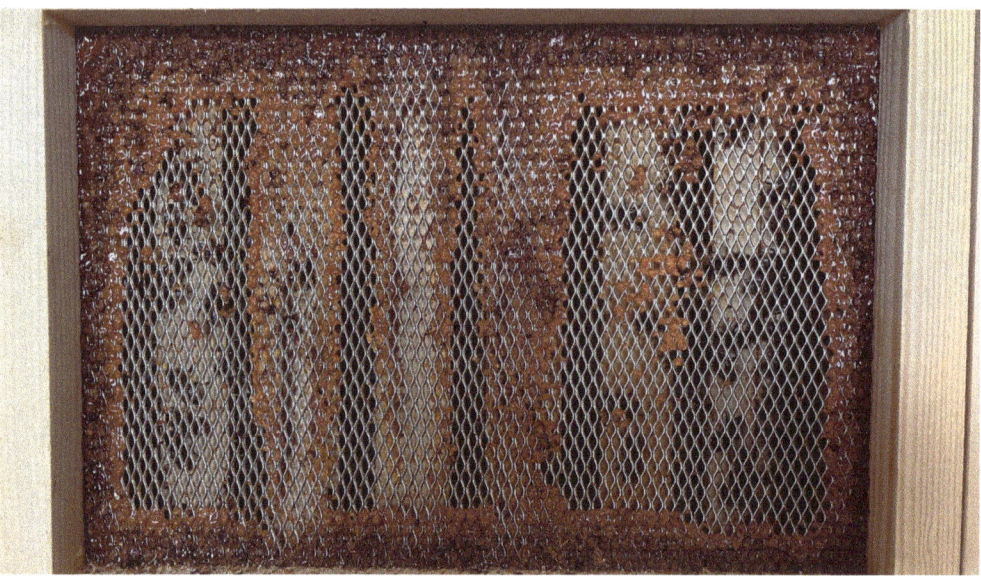

Figure 4.7.3 Week 1 propolis deposition

Figure 4.7.4 Week 4 propolis deposition

Varroa

Figure 3.5 shows that not one single varroa mite has been found on the inspection board. Varroa numbers seem to be uncommonly and exceptionally low this year. This is a very unusual situation. Last year the same colony had a depressingly high varroa drop. I wondered whether local beekeepers had similar experiences, so I posted a question on a local forum…"No, a very bad year for me, I had to take the supers off midsummer and treat" said one beekeeper.

While I am delighted to find almost no varroa in this hive, I am a little disappointed not to have the opportunity to include more pictures of varroa. Twelve months ago, I made daily observations of whole and damaged varroa. When I started planning for this study, varroa mites were a guaranteed find. It would be foolish to conclude that the increased propolis deposition has reduced the varroa load, but it is a very interesting observation all the same. A controlled experiment with replicates would be needed to conclude that the propolis deposition is the factor that has 'cured' this colony. There is however literature evidence to support this explanation as a possibility. For example, one study demonstrated that when honey bees are under stress because of varroa infestation, an increase in the number of resin collecting foragers was recorded and concluded that the collection of plant resins and its use in the hive as propolis is an example of colony defence (Pusceddu, et al., 2019).

Figure 4.7.5 shows four pictures of a 6-pore pollen structure. This looks like rosemary and other members of the Lamiaceae family, commonly known as nettle, deadnettle or sage. The flower field is full of all these, so it is difficult to narrow things down. The spiky Asteraceae pollen isn't a surprise as our flower field is full of them.

Debris from the swarm colony

In early July, my priority was to ensure the survival of this cast swarm. Feeding the swarm with a warm sugar syrup solution was easy, the weather was much more difficult. Bees were not flying and beekeepers across the country were being advised to feed their bees in June. Many local beekeepers were reporting problems with mating too. I was concerned for the survival of this small colony. The sooner a queen gets mated and starts laying eggs the better. Given good weather, the queen usually gets mated in the first two weeks of her life.

Two days after 'hiving' the swarm, I consoled myself by looking at the debris. Figure 4.7.6 shows debris dropped from the heart and the edges of the cluster. A few calendar calculations suggested that this colony still had a virgin queen and an inspection showed no eggs. The debris found nearest the edges of the cluster were mostly translucent wax flakes or chewed white wax lumps. All good signs that the nest building is happening to prepare for future brood. The column of debris on the right is

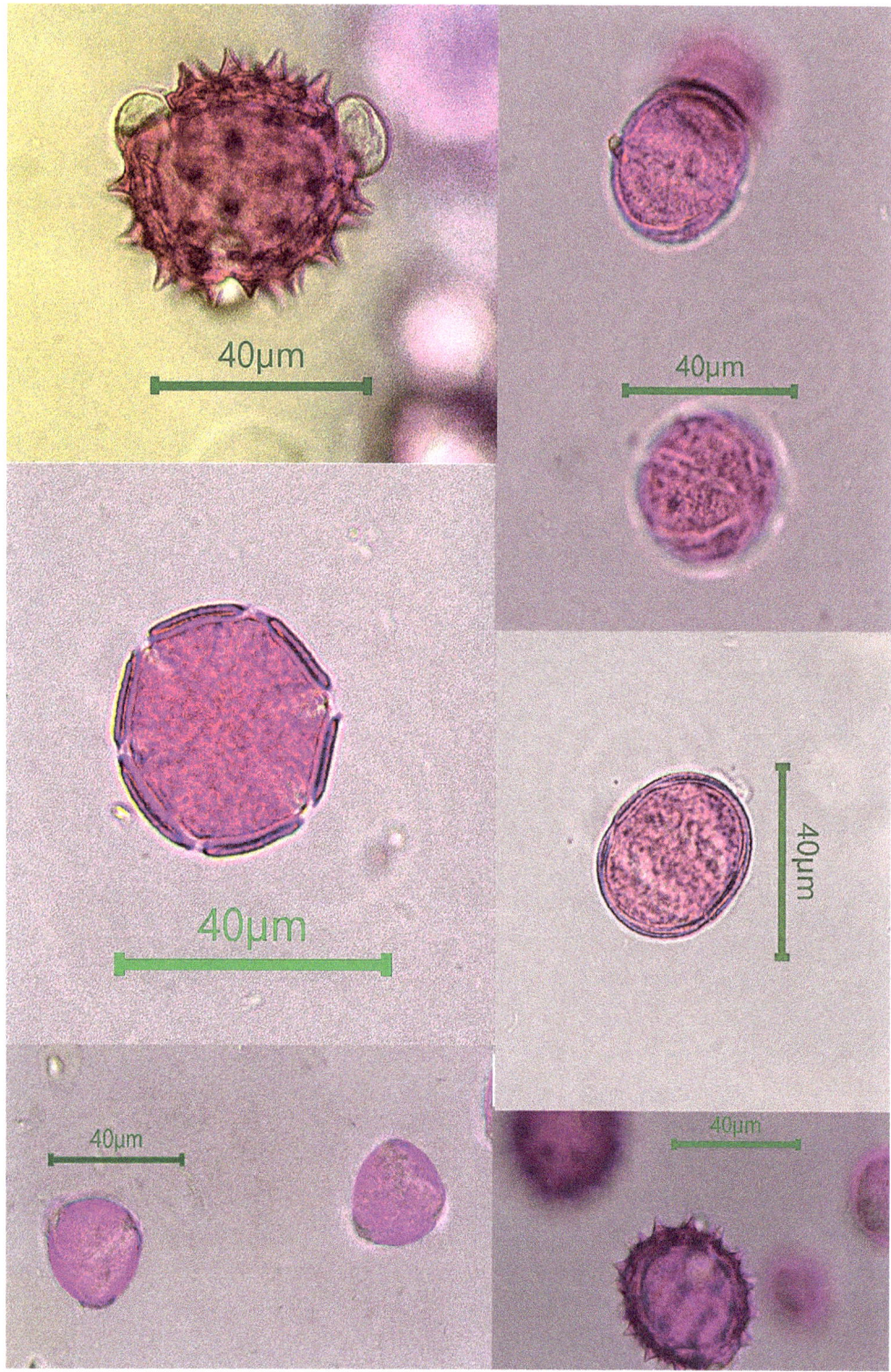

Figure 4.7.5 July pollen

from below the heart of the cluster and the column on the left is from the edges of the cluster. Near the heart of the cluster there were many wax flakes, a chewed amalgam of propolis-stained wax and many wax/propolis bee pellets containing bee hairs. The mainly white wax flakes on the outside of the cluster will have been produced by the bees in the warmest part of the colony and transported to the periphery for comb construction. All of this activity produces a colour gradient of debris from a dark chestnut brown to white. No pollen was visible in the debris at this stage.

Figure 4.7.6 Debris after two days - with a virgin queen

Figure 4.7.7 shows the images of debris deposited at 7-day intervals. This debris helps to tell a story of colony development shortly after queen mating, which is egg laying, followed by the larval and capped brood stages. The debris shown has been created during nest building, larval feeding, cell capping and colony activity that keeps a long list of variables at optimum levels for brood development: the temperature, pH, humidity, and many other variables that are critical for healthy brood development. An inspection showed capped brood and provided confirmation that this debris must have been produced by the processes described.

It is interesting that for the first 21 days the debris became darker and more spread out as the nest expanded. The change in colour of the debris is an interesting feature. I have wondered if it could be related to the amount of hormones produced by the feet of honey bees.

"When honey bees walk across a surface, their feet often deposit an attractive, oily and colourless secretion. This secretion has been shown to affect the behaviour of other workers, thus is considered to be a pheromone. This chemical has been termed 'footprint pheromone' or trail pheromone" (Collison, 2015).

This pheromone appears to assist the bee in orientation. Workers leave a trail of it at the hive entrance and on flowers, which in turn leads to more bee visits. If more foot traffic would lead to more of this oily hormone being deposited on the wax, which would help to orientate the bees, it seems plausible to suggest that this oily substance trail would become darker and more concentrated with more traffic. If this is the case, then it would seem reasonable to suggest that the colour change of comb and debris may be related to some kind of concentration gradient for this pheromone. It's always fun to wonder.

Figure 2.7.3 shows that the weight of the debris increased for this three-week period. In the fourth week there was a reduction in the debris weight and a change in debris colour. A similar trend is also visible for the number of pollen grains found in the debris. It is interesting to note that these trends, the colour changes, number of pollen lumps and the weight trend all happen over a 21-day period, this being the life cycle timescale from laying an egg to emerge. In week 4 the picture shows more fresh wax flakes in the debris as the colony continues with its nest building. It would be interesting to know if these trends are representative of all emerging colonies.

Debris 7 days after 'hiving'

The first whole pollen grains became visible shortly after the queen had been mated. Under the microscope, lots of virgin wax flakes, 'hairy' wax pellets, unidentified mites and processed wax. Figure 4.8 and 4.7.9 shows the debris produced by a virgin colony with a newly mated queen.

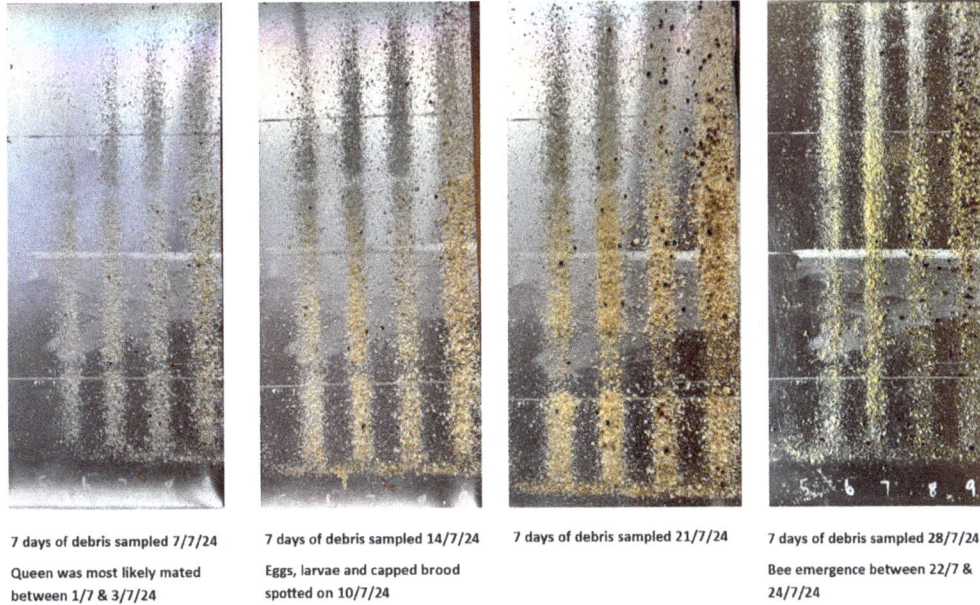

7 days of debris sampled 7/7/24
Queen was most likely mated between 1/7 & 3/7/24

7 days of debris sampled 14/7/24
Eggs, larvae and capped brood spotted on 10/7/24

7 days of debris sampled 21/7/24

7 days of debris sampled 28/7/24
Bee emergence between 22/7 & 24/7/24

Figure 4.7.7 Twenty-eight days of debris for the swarm hive

Figure 4.7.8 Debris -- post mating and pre-emergence

Figure 4.7.9 Debris produced by a virgin colony with a newly mated queen

Figure 4.7.10 and 4.7.11 show the pollen diet collected by foragers and fed to the larvae in this colony for two different weeks. A mixture of mallow, rosebay willowherb, field beans, dandelion and ling heather were found in Week 1 (figure 4.7.10).

Figure 4.7.11 shows that by week 4 the diet had changed showing no rosebay willow herb, marrow or field beans, but lots of phacelia (top right) and Asteraceae (right middle)

Comparing these pollen finds with the pollen found in the debris from hive 1 seems to show two different colonies in the same week feeding on completely different forage. Different amounts and different colours of pollen are present for two colonies less than a metre apart. Why would two colonies in the same apiary forage on different pollen? A quick glance at the colour of pollen grains on the inspection board of both colonies showed that the bees were indeed foraging on different plants. Could the bees be making dietary 'choices' about the kinds of pollen collected and could this choice be related in some way to the development stage of the colony?

Figure 4.7.12 shows a surprise find growing in the debris of this new colony. A carpet of white fungal mycelium growing on pollen. This growth happened in just a few days. I instinctively removed and cleaned the inspection board. My concern was the upward movement of pathogens into the colony. I also checked inside the brood nest and there was no sign of this growth. It seems that bees had this under control. My best guess is that fungal growth became possible because of the ventilation screen, trapping the debris below. It seems likely that without this screen the bees will have processed these pollen pellets into bee bread and that this kind of growth would not have happened. It seems that unchecked growth of this fungus will inevitably become a food source for a host of other organisms.

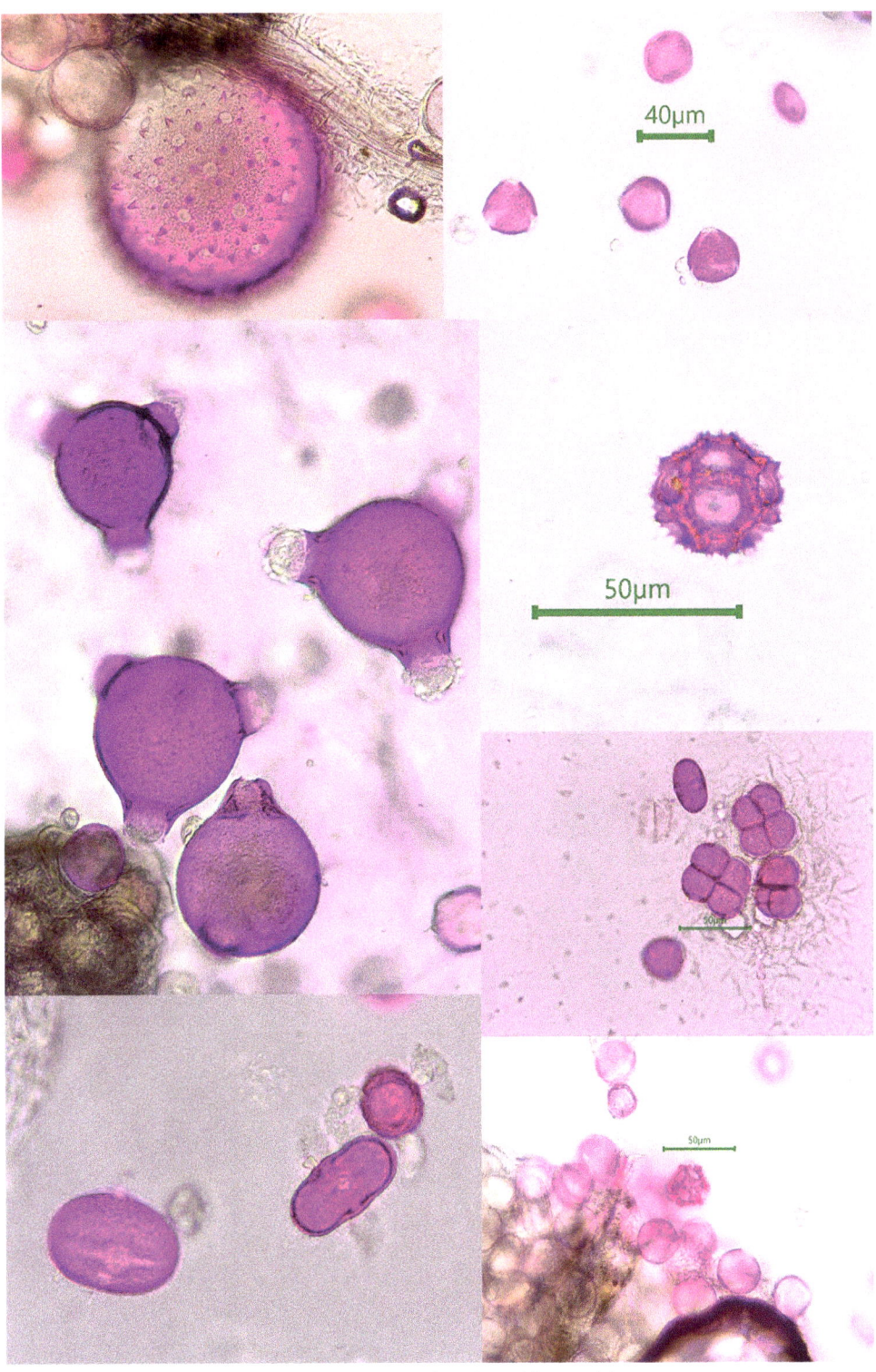

Figure 4.7.10 pollen found in week 1

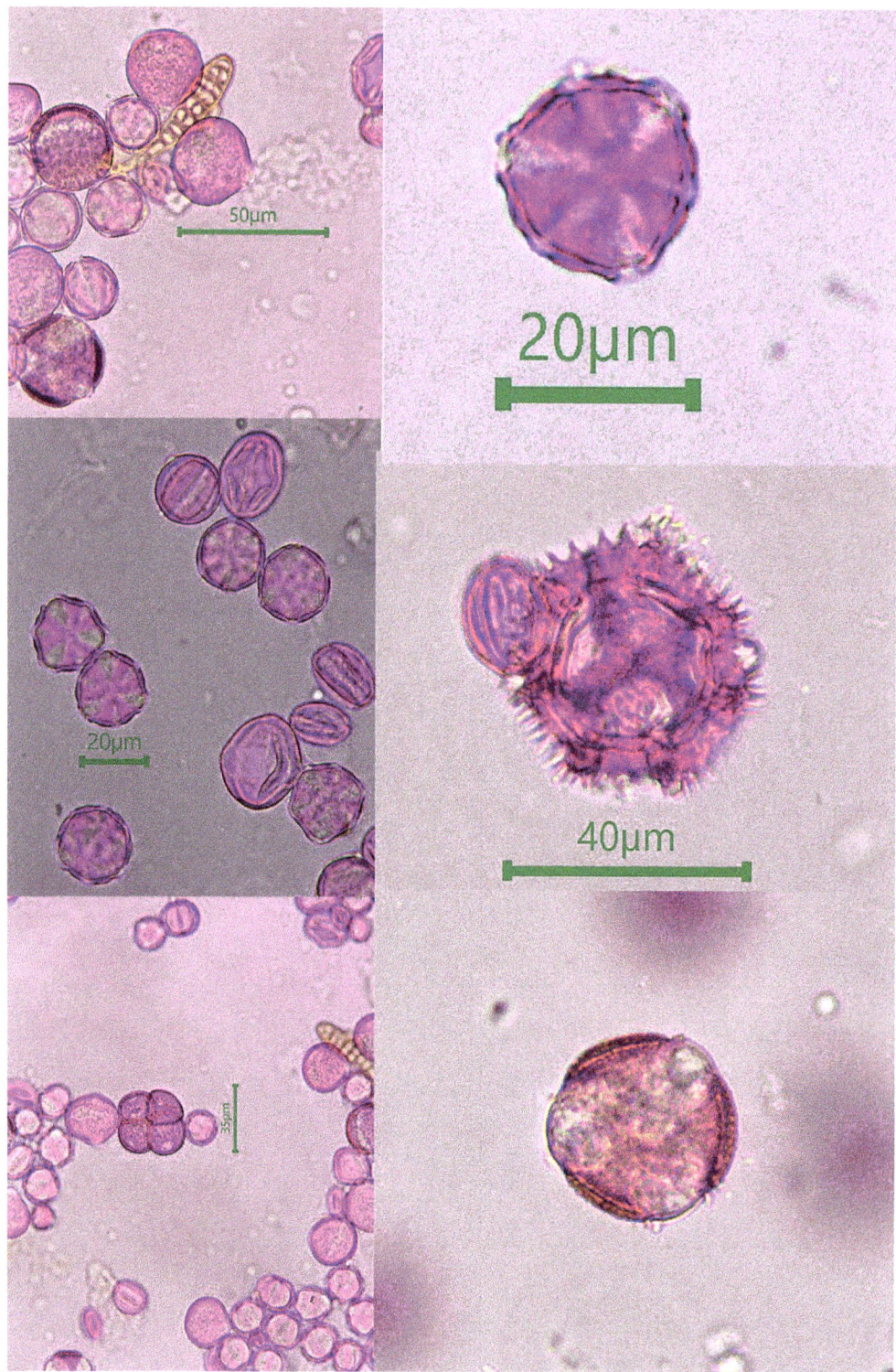

Figure 4.7.11 pollen found in week 4

Figure 4.7.12 Fungal mycelium in the debris

This find also made me think about how we are trained to be extra vigilant against sneaky microbial attacks. Constantly on the lookout for bacteria, fungi and viruses that can cause harm.

It's a fact that there has been a research bias toward organisms that can cause harm. How many of us can talk as knowledgeably about Lactobacillus kunkeei or Saccharomyces and their role in ensuring the survival of our honey bee colonies or describe the real threats there are to the survival of these essential microbes, some closer to home than we may imagine. The work of these microorganisms preserves pollen stores through a process of fermentation to produce bee bread, an essential food source to get bees through the winter produced by a similar process to making beer, cheese and other fermented foods. The honey bees' health and survival are dependent on this processed food.

The process starts with worker bees pushing pollen pellets into cells. The bees add salivary secretions and honey. When the cell looks full the bee uses its head to push more pollen into the cell until it is full, before adding a final layer of honey to exclude air. The microbes do the rest. The process consists of bacterial fermentation followed by fungal fermentation. In the bacteria stage, lactic acid is produced by a species called Lactobacillus kunkeei. The production of lactic acid lowers the pH and sets up the perfect environment for the next stage involving fungi belonging to a family of Saccharomyces. It is the same fungal yeast that is used by humans to make beer and sourdough. These microbes preserve the bee bread for feeding by nurse bees to young larvae after three days of age. With this in mind, we should be concerned about how fungicides commonly used in agriculture have been found to alter and reduce fungal communities in honey bee colonies. Research also suggests that beekeepers may be decreasing the species richness and abundance of microbes in the hive through the use of miticides commonly used to treat varroa (Brooke, 2022) (Lopez-Uribe & Lawrence, 2021).

Unprocessed pollen has a short shelf life, just like the plants that they were destined to become. Perhaps finding this growth is a good sign for my bees, suggesting that fungicides have not been applied at lethal levels in local fields. Finding microbial growth in the debris has made me realise that I do not understand enough about the honey bee microbiome. I need to find out more about the nine clusters of bacterial species that live in the gut of the honey bee (Kwong & Moran, 2016) and also to better understand how fungi also influences bee behaviour and health. Studies have shown that yeasts can affect bee foraging, honey bee development and nutrition, and that fungal communities associated with honey bees change throughout a season (Rutkowski, et al., 2023). Understanding the role of these beneficial microbes is surely as important for bee health as understanding the ones that can do harm. Perhaps the debris produced by a colony has the potential to answer some of these questions. I would love to have incorporated DNA analysis into this study and to be able to identify the microbial species in the debris each month, but it's an expensive option for an independent researcher. I have kept samples from each month in the freezer just in case this becomes an opportunity in the future.

Emergence of the first workers

Calendar calculations suggested that the emergence of the first workers would be near the 23rd July. Around this time, I stumbled upon a fantastic photograph posted on Instagram by 'Sallythebeekeeper'. Sally kindly gave permission for this photograph to be included (see figure 4.7.13). The photograph shows an emerging bee using its mandibles to nibble around the edge of the wax cap, an action like opening a can of beans that removes the 'wax lid' from the cell body. It is easy to imagine how this process results in different sizes of debris, from fine particles from the cell structure to whole brood caps. Figure 4.7.14 shows some of the debris after emergence.

Figure 4.7.13 An emerging bee (Sally L. Ewen)

Figure 4.7.14 Debris after emergence.

4.8 August

Drone culling

Honey bee debris collected between 1st and 28th August

Minimum temperature 5°C

Maximum temperature 25°C

Mean temperature 15°C

Figure 3.1 shows a 46% reduction in debris compared with the previous month. This trend can easily be seen by looking at the inspection tray (figure 4.8.1). This downward trend represents a change and reduction in colony activity. It is expected that this trend will continue until activity plateaus producing a corresponding plateauing of debris deposition through the winter.

Pearly white skin-like structures were found daily over a two-week period at the beginning of August. With the naked eye, it is possible to see that these are the cannibalised parts of pupae See figure 4.8.2). The microscope helped to identify these fragments of exoskeleton. Figure 4.8.3 shows a small sample of these body parts.

I assumed that the pupae were killed, the inner parts eaten and that external cuticle fragments discarded in the debris. The interesting question is why and why now? What is going on in the colony?

I started looking for answers. Matthew Richardson (President of The Scottish Beekeepers Association) asked if I was finding any whole antennae. Matthew's thinking was that antennae could be used to determine the sex of the pupae. A bee's antennae are divided into segments. Females have 10 and males have 11 segments. Figure 4.8.4 shows a male antennae found in the debris.

Over a two week period 23 antennae were found, 20 with 11 segments, a mixture of drone and worker bees, but mainly drones. The translucent quality of the pupal antennae shows the outer surface and inner structure. The outer surface is covered with different types of receptors, each type having a special purpose. Estimates vary, but each worker antenna has roughly 3,000 chemoreceptors, whereas a queen's antenna has only about 1,600. But drones, whose job is to find virgins queens in mid-air, have an estimated 300,000 chemoreceptors.

A plausible explanation for the drone antennae could be the culling of males and their removal. It is a common sight in the autumn to see males being thrown out of the entrance. The males have fulfilled their reproductive roles for this year and space and resources can be better used. Their sperm however remains viable and is stored by the queen inside her spermatheca until the following season.

Figure 4.8.1 August debris [WHOLE PAGE]

Figure 4.8.2 Cannibalised pupae parts collected from the debris

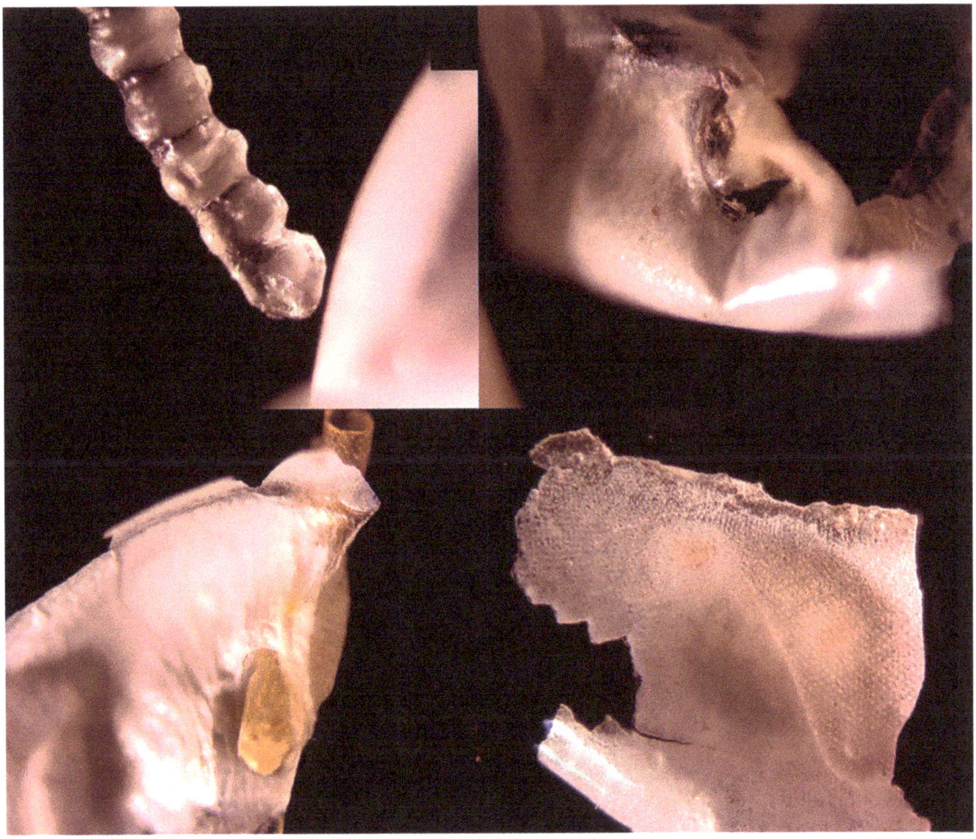

Figure 4.8.3 Cuticle parts

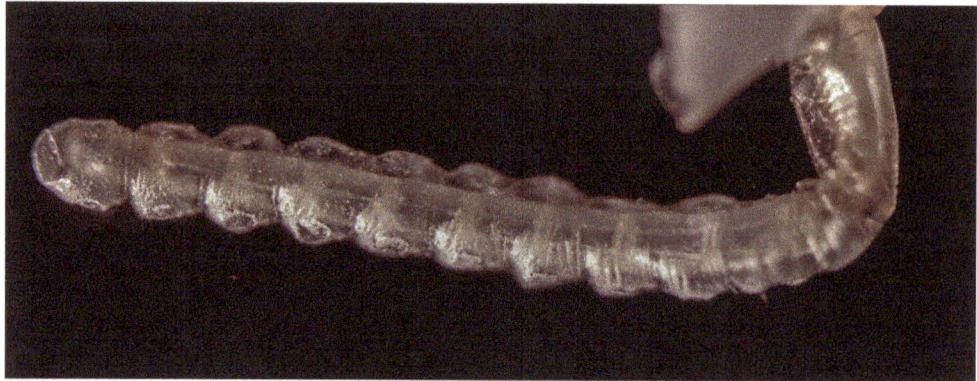

Figure 4.8.4 A typical antennae found in the debris.

Another explanation could be varroa sensitive hygienic behaviour where worker bees remove mite infested pupae and cannibalise the pupae. This behaviour is well documented in honey bee research and is generally regarded as a good thing. However, some research suggests that there may be negative consequences associated with this kind of varroa sensitive hygiene. Research has shown that cannibalisation of pupae infected with deformed wing virus can become a transmission route for infection, allowing deformed wing virus to pass from bee to bee during mouth-to-mouth feeding (Posada-Florez, et al., 2021).

Bee's antennae are remarkable structures, with different segments for different sensors; temperature, humidity, carbon dioxide, gravity, shape, vibrations, pheromones, sugar concentrations and taste. Figure 4.8.4 shows the inner nerve that transmits a seemingly endless supply of data to the bee's brain. There is a sense of awe and wonder amongst humans struggling to understand what the bees do with all this information.

In the first few months of this project, lots of adult bee exoskeleton parts were found and the emergence of pupal cuticle fragments in August raises the issue of cannibalism again. Cannibalism is common in the group of insects that include social insects. This large order of insects called Hymenoptera contains many examples of cannibalism in ants, bees, wasps, and termites. It has been recorded throughout the growth and development of social organisation of different kinds of colonies (Posada-Florez, et al., 2021). Research has shown that that bees recognise the status of brood cells by pheromone signals released by the larvae and pupae and can use these signals as a cue for cannibalistic behaviour (Santomauro, et al., 2004). It is assumed that that cannibalism occurs for many reasons, including food shortages, hygienic behaviour, disease, colony disturbance, policing of haploid egg laying, pollen scarcity, unbalanced nutrition, regulation of the population, and doubtless many more.

This bee story would not be complete without mentioning the flying hooligans of the insect world! Unfortunately, we have been poorly educated and misinformed about the relationships between bees and wasps and their roles as stewards of the environment. To do both a favour, we need to direct the narrative away from wasps being on the dark side versus flower-loving bees on the light side. Both are anthropomorphic generalisations that profoundly shape how people understand their lives.

The Wasp Professor says "wasps are the ancestors of all bees" (Sumner, 2022). Time machines that can help us better understand honey bees. It's a fascinating topic, especially considering evolutionary changes in mandible morphology, nutrition and debris. The ancestors of bees have changed their diets many times. The first wasp was a vegetarian, others developed a taste for meat and everything in between. Intriguingly the switch from vegetarian to meat eater did not result in significant changes to the digestive system. Scientists think that the complex diet of pollen, nectar and plant tissue could easily make the transition to carnivorous diet possible (Sumner 2024). It is pure speculation, but could this evolutionary lineage demonstrate that honey bees are capable of utilising the nutrients of both plants and animals? However, comparing the mandibles of the common wasp, a honey bee would be at a distinct disadvantage, not having three frontal teeth on each mandible as found on Vespa vulgaris. No doubt honey bees would do a better job at dealing with pests if they had sharper teeth. Perhaps the increase in invasive pests such as varroa, the Asian hornet, small hive beetle etc are creating a selection pressure for toothy mandibles. It's an intriguing thought that in the longer term, evolution may provide a solution for these pesky troublemakers.

But how much damage can a bee really do? Figure 4.8.5 helps to illustrate. It shows chewed newspaper bee bite marks created while uniting two colonies over a 24-hour period. To help illustrate the destructive power of mandibles, imagine a 12mm tall bee standing on its back legs. I'm almost 150 times taller than the bee at around 1800mm. In less than 12 hours these bees chewed a hole in a newspaper that was around 0.1 mm thick and 80mm wide. Multiplying these values by 150 gives an equivalent human scaled value. In human terms the thickness of this chewed newspaper is around 15 mm deep and 12 metres wide. That's like me eating the entire downstairs carpet including the underlay in my home. It seems that bees can create a lot of damage.

Figure 4.8.5 Bee bite marks

Honey bees are not 'flower loving vegetarians' as often portrayed, only consuming pollen and nectar. While they may not hunt quarry for their nutritional needs they are superbly adapted to defend, and in the act of defence they bite, chew, tear and manipulate all types of materials, including tiny animals and other bees. While 'meat' is off the menu as the key part of the honey bee diet, the evidence is that they can and do benefit from occasional animal nutrition under particular circumstances. All this activity leaves a debris trail. In the first few months of this study, exoskeleton parts of adult bees were a common find in the winter, much less so in the warmer months and in August they have become common again

In the case of male pupal fragments, it seems plausible that the debris is showing a kind of sexual cannibalism and/or varroa sensitive hygiene. The males focus on ensuring the paternity of future generations has been achieved. The bee sperm continues to 'live on' as remarkable genetic survival machines stored in the queen's spermatheca. The debris shows that the females remove and break up the male pupae and that only male cuticle parts can be found in the debris. The assumption is that the inner parts of the pupae become part of the nutritional diet for the colony and that the space occupied by drones becomes a food storage area in preparation for the winter.

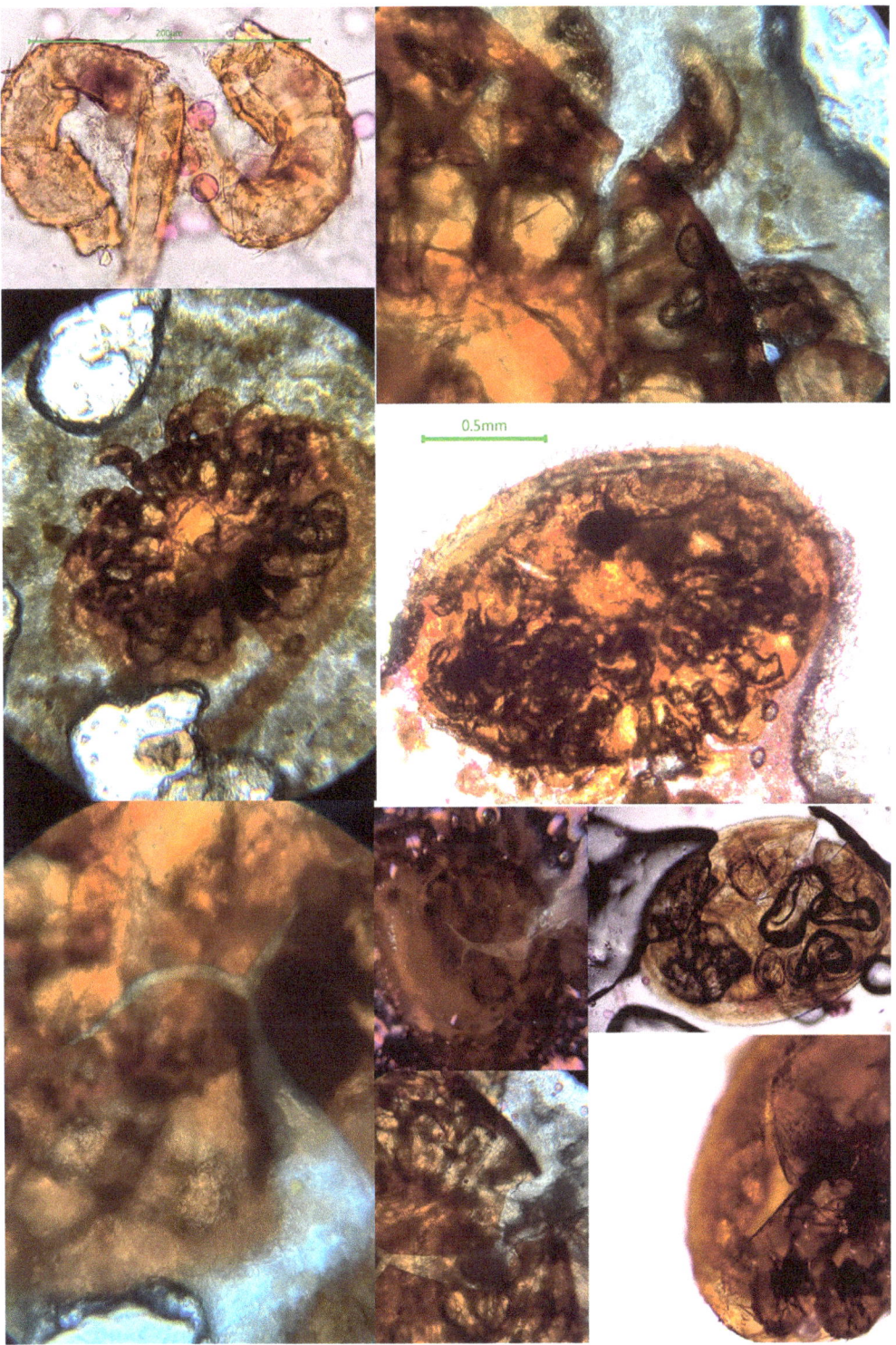

Figure 4.8.6 Varroa bite marks shown only by microscopy

We really shouldn't underestimate the power of bee mandibles. These are extremely strong hinged structures with a cutting edge. The first activity of all honey bees involves using these cutting tools to open the brood cap. Each mandible is operated by two strong muscles, the second largest muscles in a bee's body, after the flight muscles. Imagine if the human jaw was powered by our hip muscles to know their incredible strength. We would literally be able to eat a horse!

In addition to this muscular power, there are glandular secretions produced near the biting edges of each mandible This secretion produced in the cheek of the honey bee is known to cause paralysis of small animals like varroa and wax moths and also act as an alarm-like function for other bees to join in. Add to these mouthparts, 12 additional 'biting' claws, two for each foot and we really shouldn't be deceived by the regular 'bee-washing' that distorts the truth by depicting bees as friendly vegetarians. These cutting/killing tools are an essential part of a colony's survival.

Figure 3.5 shows that a single whole mite was found in the debris over this 28 day period. However, through the process of microscopy the picture became more complicated. Encapsulated in the debris I started finding missing legs, feet and obvious bite marks on mites. Varroa could easily be found that had had their legs nibbled and outer shells broken. None of the mites in the pictures were visible with the naked eye, but microscopy shows many more. These were only revealed by dissolving wax debris and using 100 x magnification, raising a few questions about the validity of a mite count for monitoring varroa. Using the recommended method of counting mites on the inspection board I would have recorded only 1, using microscopy there were 100s of mite fragments. Figure 4.8.6 shows a few examples of these damaged mites

It is also interesting to speculate reasons why there could have been an increase in the number of body hairs found in the debris. Figure 3.3 shows a higher number of body hairs than any other time in this study - the debris had become much hairier. Figure 4.8.7 shows a few examples of these hairy finds

Propolis under the microscope

In previous months I discussed the many virtues of propolis and the idea that there may be human health benefits. It seems that there may also be potential for oral hygiene, healing minor wounds and treating cold sores. A recent systematic review of the clinical evidence states "that propolis-based mouthwashes showed promising clinical outcomes in reducing plaque and inflammation. However, it is recommended that to conduct more rigorous trials with patient reported outcomes" (Ballouk, et al., 2025)

With a developing sense of awe and wonder for propolis, I decided to put a lump under the microscope. I prepared the propolis in two different ways. The first involved picking off small lumps of propolis and looking at these samples in an unprocessed state, The second method involved dissolving the propolis in alcohol before viewing using a microscope (figure 4.8.8 and 4.8.9)

Figure 4.8.7 Hairy finds

Figure 4.8.8 Propolis under the microscope

Figure 4.8.9 Propolis samples dissolved in alcohol

I really wasn't expecting to find anything other than different colours of resin. Imagine my surprise to find lots of very small fragments of exoskeleton. I know that when bee boxes are added and removed during hive inspections, sometimes bees get trapped and I wondered if it could be an explanation. I decided to start looking for propolis in places where it had been freshly deposited such as in areas where there was no possibility of beekeeping introducing bee parts. The experiment was repeated using twelve samples from two different sources of propolis. The first source was from a large lump of propolis deposited by the bees near the entrance and the second source was removed from the ventilation screen at the rear of the AZ hive. These twelve samples were studied in a dry state and after dissolving the resin using alcohol as described. Figure 4.8.10 shows a small selection of these finds in the propolis in a dry state and figure 4.8.11 shows a selection of finds after dissolving the resin.

I can say with confidence that the propolis sampled from my apiary contains many thousands of tiny fragments of exoskeletons (between 5μm and 100μm). Why do bees pack propolis with so many tiny fragments of bee exoskeleton and then squash it together in one lump at the entrance or on the ventilation screen? Most of the exoskeleton parts are hairless and there is no soft bee tissue to be found. Are these fragments a kind of aggregate mixed in the propolis for construction and sealing gaps or is there another explanation? Moments of discovery like these are so exciting. In previous years, a few ex-students from Eastern Europe handed me handwritten notes from family members asking if they could buy propolis to make a winter tincture. I have often thought of making my own, but the idea of having bee parts in a tincture is now less appealing - maybe I'm being squeamish and should have more faith in the antimicrobial properties of this bee product.

In August the most common pollen was phacelia. Figure 4.8.12 shows these pumpkin shapes at just over 20 microns in diameter. When viewed from their poles they look round in shape and from the equator more oval in shape. At first glance they look like they have 6 pores, but on closer inspection they have 3 furrow type apertures, which extend almost to the poles and another 3 furrow like features without apertures that do not extend quite so far. The rectangular four pore structure is Himalayan balsam which grows in moist and semi-shaded damp places including riverbanks, and thin woodlands. Field bean pollen can also be found along with rosebay willowherb has a distinctive large shape and the pollen load is easy to identify due it being blue. Its pollen has a swollen round shape with three pore-like apertures which stand proud of its smooth surface (middle left).

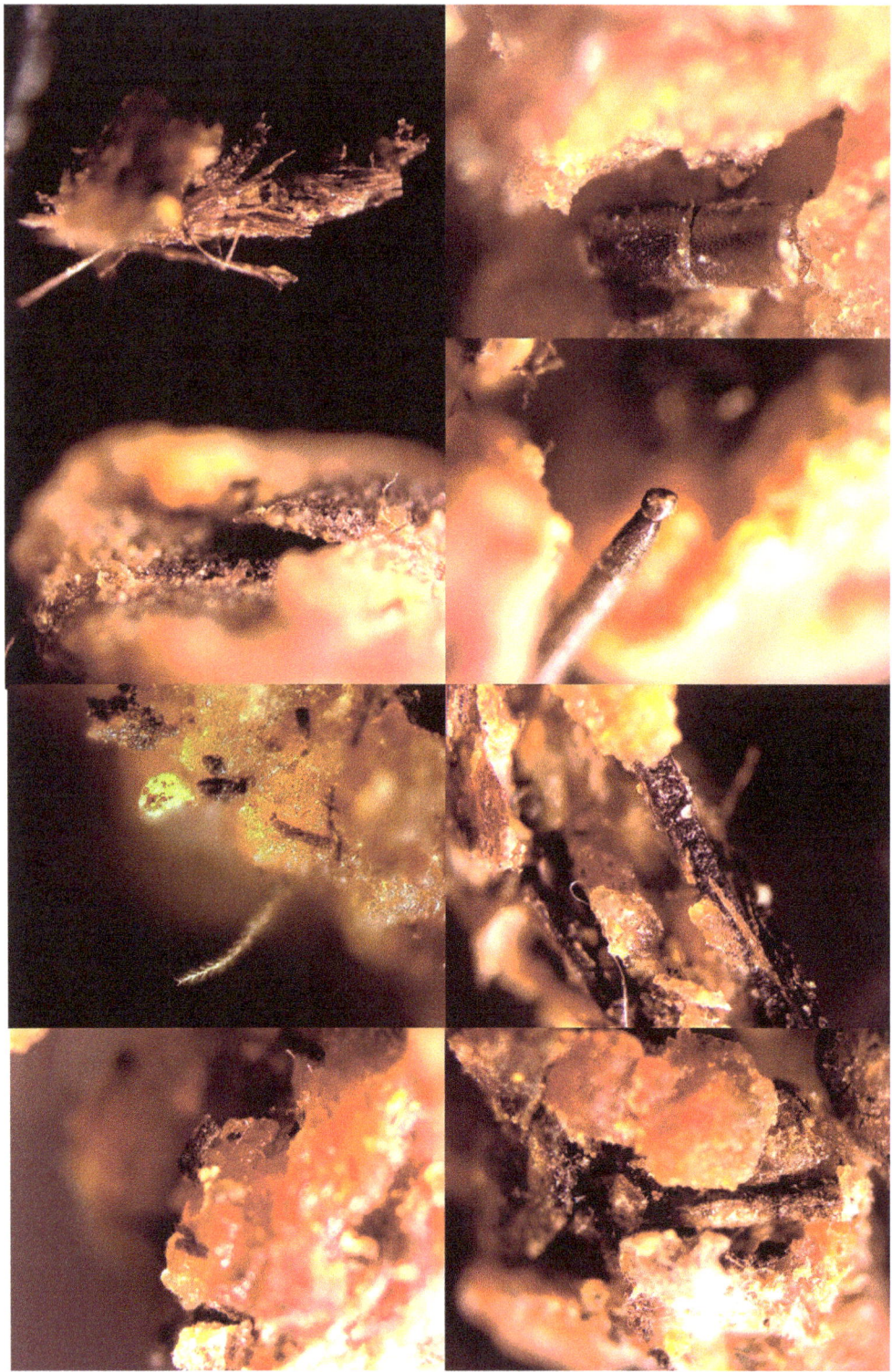

Figure 4.8.10 Bee body parts in dry propolis

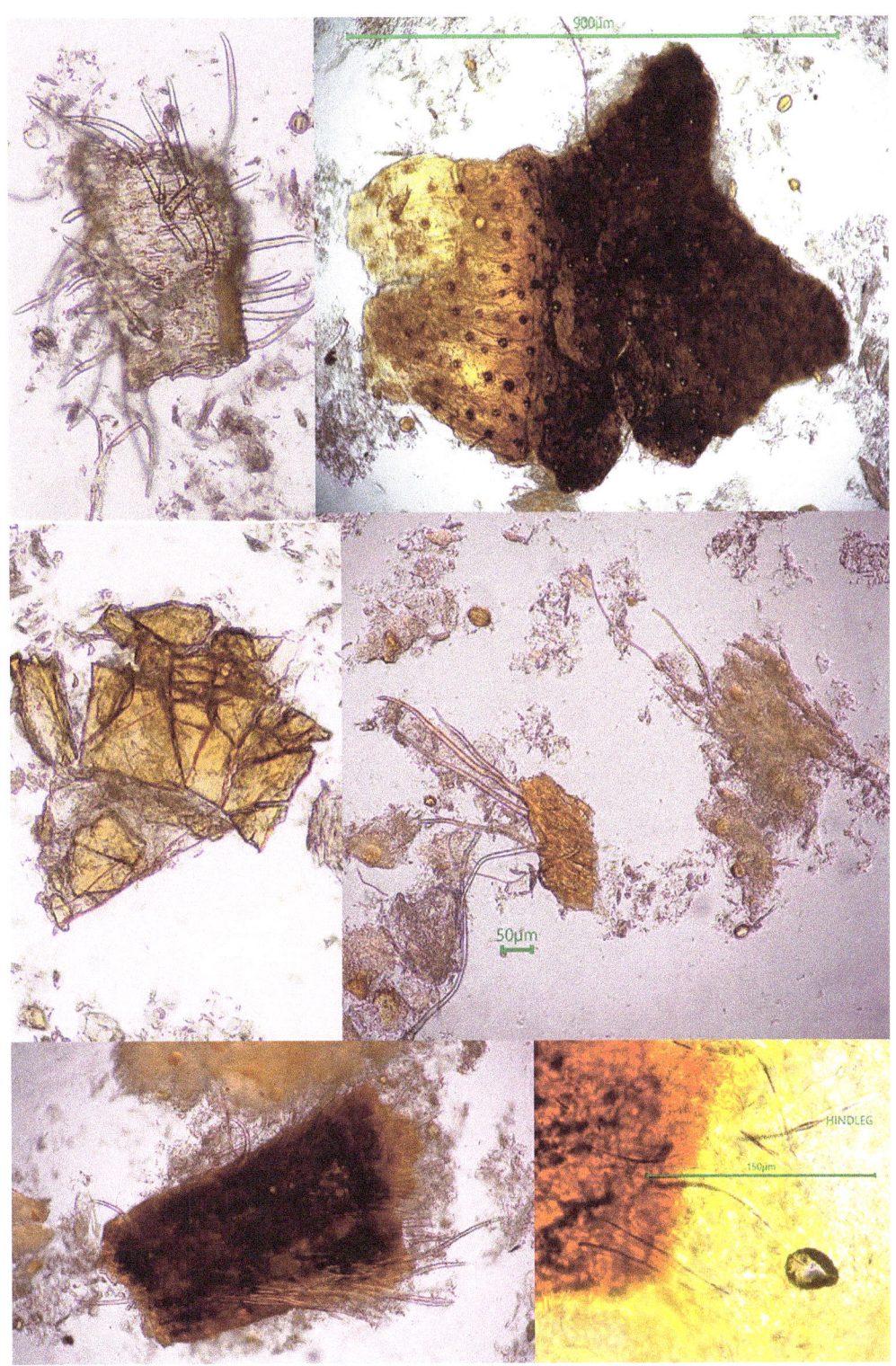

Figure 4.8.11 Bee parts in dissolved propolis

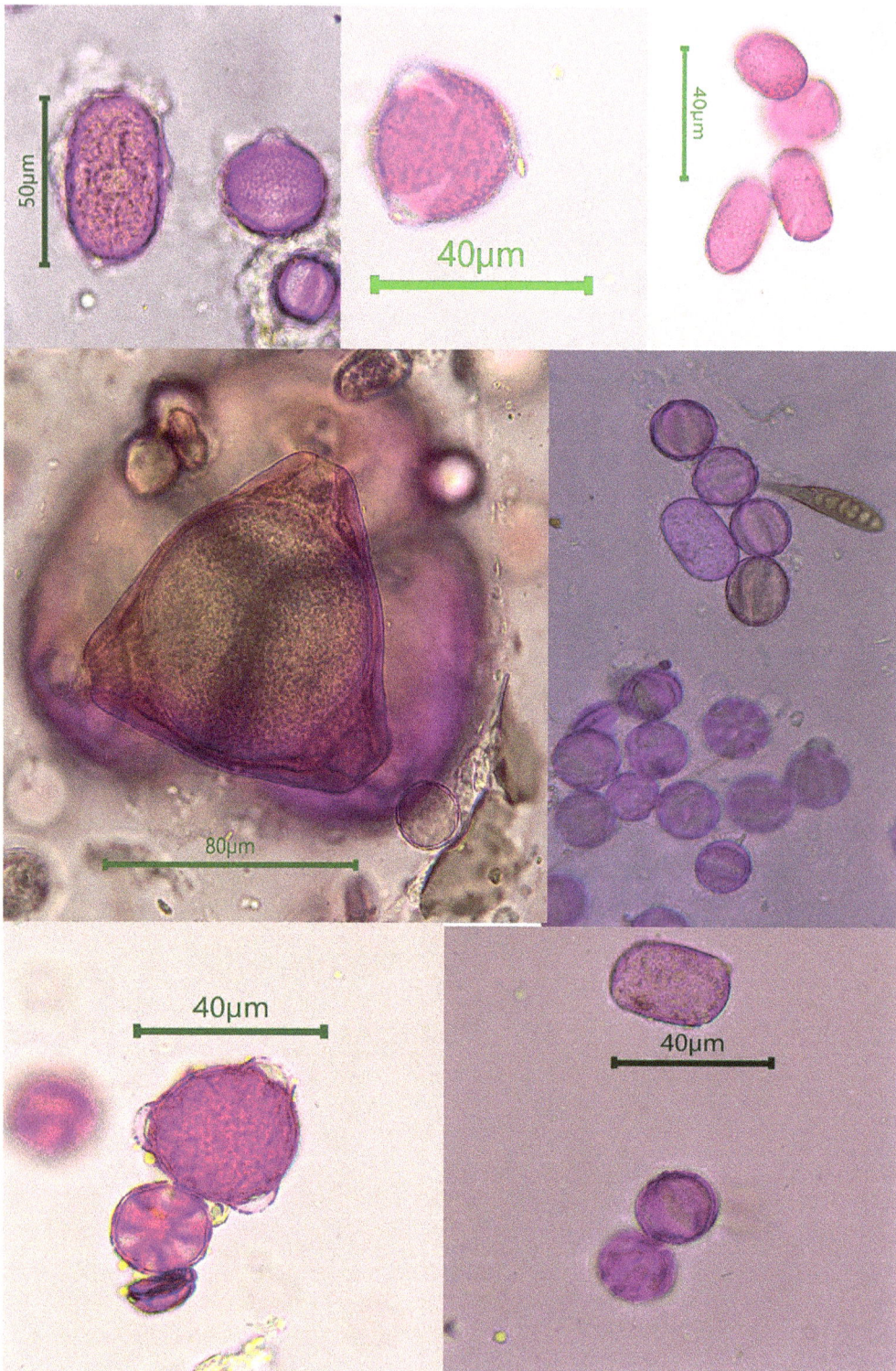

Figure 4.8.12 August pollen

After completing the debris analysis, I extracted this year's honey bounty before replacing the inspection board on 1st September. This single hive in one season produced 18 full AZ frames of capped honey. Each frame yields 2.1kg of honey (figure 4.8.13) or in total 37.8Kg of food stores. I would be delighted with this crop in any season, but especially when the weather has been so challenging. I gave the local farmer a few jars of this liquid sunshine and thanked him for planting phacelia as a cover crop. I take 10 frames for processing and leave 16.8 kg of honey for the bees. I hope that this will be enough to see the bees through the winter without any artificial feeding and will also help the bees get off to the best of starts in the next season.

In my opinion, deciding how much honey to leave for the bees to overwinter is one of the most important decisions a beekeeper can make. I am grateful to Ann Chilcott for her recent blog post "Overwintering on Honey: Is it really any better for bees than artificial feed?", where she presents a review of the literature. Ann describes the real threat of honey bee malnutrition and explains why some of the most serious consequences of malnutrition are impaired immune systems and increased susceptibility to pesticides. This literature review concludes that bees fed honey have 'better' gene expression than bees fed artificial food (Chilcott, 2025).

Extracting the honey gave plenty time to ponder possible relationships between the debris trail and the honey yield. This one hive produced 224 grams of debris to date and yielded 37.8 Kg of honey, with the debris composition as described. It is somewhat satisfying to know that the recorded debris trail can lead to this amount of honey and some sense of achievement to now have some baseline data for when I next look at the debris in the future.

Figure 4.8.13 A summer honey crop of 37.8 Kg

2.9 September

Shorter days

Honey bee debris collected between 1st and 28th September

Minimum temperature 1°C

Maximum temperature 22°C

Mean temperature 12°C

Figure 3.1 shows a 71% reduction in the weight of debris deposited compared with August, which seems like a significant reduction in colony activity, with a deposition rate most like March. Figure 4.9.1 and the heat maps also demonstrate other similarities with March, showing the colony now occupying a similar space in the hive. These similarities have made me consider the influence of day length on the activity of the colony.

My apiary is located at a latitude of 55.7 degrees, a place that has experienced the same period of day length since the formation of our solar system. It is the axial tilt of the earth relative to its orbit that produces an annual cycle of day length and the seasons. Day length is the most regular change ever known at the place where my hive sits today. Figure 4.9.2 shows the constant of day length changes each month at my apiary.

Interactions of day length and temperature are the key cues for change in many animals and plants, for example, when a plant produces flowers, when an animal migrates, grows fur, hairs, building nests, hibernates and the display of sexual behaviours. They are all linked to the seasonal rhythm of day and night. Honey bees are not like animals such as hedgehogs which hibernate to lower their body temperatures to match their surroundings and enter a state of torpor. The strategy used by honey bees and many other insects is called diapause. This form of dormancy includes behavioural and physiological responses that govern population size, maintain a constant temperature and many other metabolic processes.

One study interested in understanding the interactions of day length and temperature found that experiencing cold conditions alone did not affect brood onset, but day length altered the impact of a higher ambient temperature on brood rearing activity (Nurnberger, et al., 2018). These results showed that daylength clearly had a role in modulating brood rearing activity. Interactions between temperature and day length synchronise the mutualistic partnership between bees and flowers. Research has implicated a circadian clock in key aspects of honey bee foraging for flower rewards. These include anticipation, timing of visits to flowers, time compensated compass orientation (Bloch, et al., 2017) but is more difficult to determine whether these oscillations are driven by time, day length or some other factor.

Figure 4.9.1 September debris

Several receptors have been found in the brains of insects, in compound eyes and larval stemmata that are believed to control responses to day length. Two groups of molecules believed to be involved include a blue light sensitive receptor and opsin proteins conjugated with vitamin A molecules that absorb light at different wavelengths (Saunders, 2012). One study using a parasitic wasp, a relative of the honey bee, found that a day length of between 12 and 14 hours of daylight was found to be a cue for a physiological switch that determined whether certain genes are activated to cause this insect to enter a winter (diapause) or summer (non-diapause state) (Saunders, 2012).

It's perhaps not surprising that the trend of debris deposition is like the day length trend shown by figure 3.1. Debris deposition is at its lowest just below this cue for the wasp to enter its winter state. Could it be that the same period of day length is also critical for bees? The rate of debris deposition started to increase slowly after a day length of 10.6 hours. With a longer day length of 12.9 hours the deposition of debris increased at a rapid rate until some point between 28th May and 28th June. The summer solstice falls between these dates, with a yearly maximum 18 hours of daylight. After this point, the rate of debris deposition decreases until it levels out again in September, a month with below 12 hours of daylight.

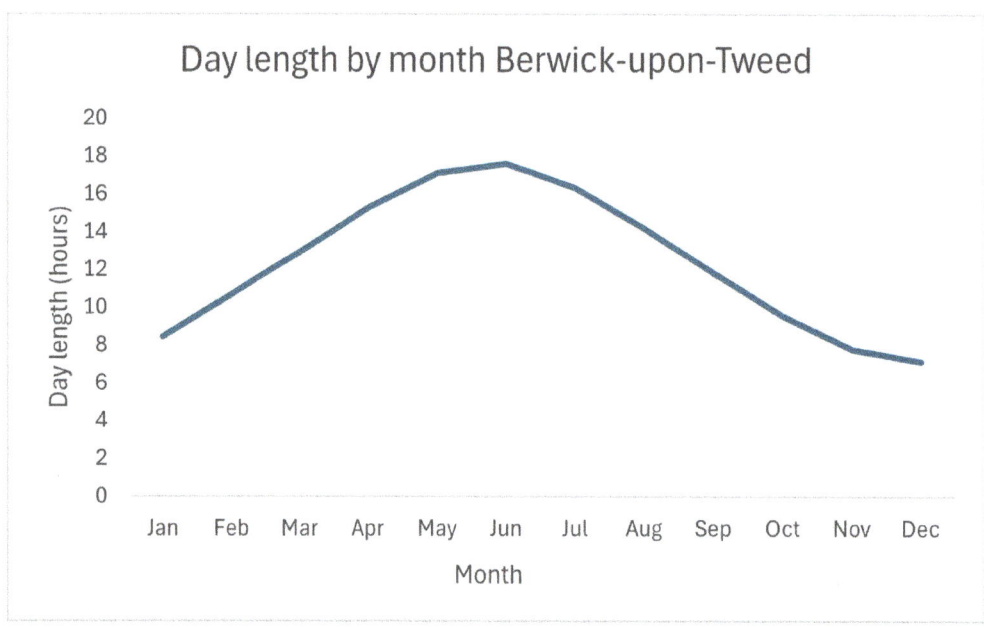

Figure 4.9.2 Day length by month for my apiary

Do honey bees moult?

Figure 3.3 shows that in both August and September there have been over 3000 body hairs per gram of debris compared with around 300 bee hairs per gram for May and June, a ten-fold increase in hair deposition in the autumn. Microscopic observations show sample hairiness to be one of the most striking differences for these two time periods. This was surprising because it seems nonsensical that a smaller population of bees will lose more hair at a time when it seems that hair is needed for insulation. My assumption had been that more hair would be lost during the summer months because of the amount of grooming that happens due to foraging.

Bee hairs, like a bird's feathers, do not last forever. They will break down by mechanical deterioration from flight, foraging and constant grooming will all lead to losses. Many other animals have a twice-yearly cycle of replacing old, worn-out hair with new, but do bees do something similar?

With September temperatures of between 1°C and 22°C, this month has already created challenges for my honey bees, which need to keep warm enough to move food stores around the hive, heat up the brood nest to replace workers, care for the queen, maintain metabolic processes and remove waste. It's a problem for individual bees who have a large surface area in relation to their volume, meaning that they lose heat very quickly. A bee must not allow itself to get too cold or it will enter a chill coma and die. To get around this problem the bees work as a superorganism by clustering together to reduce the overall surface area to volume available for heat loss.

This cluster has two parts, a dense outer mantle and a looser inner core. In the outer part, mantle bees stick their hairy thoraxes together to form a thermal shield, with their faces pointing inward towards the centre of the cluster. At the core, the bees are loosely packed and can freely move around. Near the core are heater bees that vibrate their wing muscles to generate heat for the brood nest and care for the queen. The temperature is at its highest at the centre (27°C-35°C) and decreases towards the outer mantle (10°C-15°C) (Minnaud 2024). Hairs are understood to be a very important part of climatic control.

Some years ago, a few of my students were interested in studying heat transfer in a colony so they built their own temperature sensing equipment using a Raspberry Pi computer. Together we learned some basic programming skills and added several temperature sensors inside and outside the core of the cluster. Our DIY school research demonstrated how effective the mantle was at preventing heat loss. In the winter it wasn't uncommon to find sub-zero temperatures 10 cm outside the mantle with a constant 30 degrees inside the core. The remarkable ability of bees to prevent heat loss created brilliant discussion amongst students about the value of human efforts to insulate the hive roof. It seems that we constantly underestimate the ability of bees to look after themselves. The cluster expands and contracts in relation to temperature, like a balloon, inflating and deflating as the temperature changes. It seems feasible that in warmer weather, bees will untangle their hairy legs and thoraxes creating gaps in the mantle. Also, in colder spells, muscular bee legs will contract, pulling and compressing thorax hair tightly together in the mantle. It could be through gaps in this mantle that the bees control the diffusion of gases, temperature and water vapour in and out of the brood core in response to metabolic needs. One of the options is that channels are created when hairy bees release their grip to control the temperature and diffusion of water vapour and gases in and out from the inner core. Also, could these hairy channels provide the doorways for bees to make quick forays to collect food from the colony's stores.

If tight clustering at the mantle prevents heat loss, It seems inevitable that a tight cluster will reduce the amount of debris that can 'accidentally' pass through the mantle onto the ventilation screen. It's likely that the outer mantle of the cluster will itself become a debris trap and that there will be more opportunities for honey bees from within the core to catch and recycle debris produced in the inner core. While it seems very possible, I have no evidence to support this view. Anecdotally the winter debris does seem to contain finer particles than the summer debris, but research would be needed to establish if this were true.

The study of debris deposited in September created much thought about the interactions between temperature and day length and when beekeepers choose to carry out their spring inspections. The National Bee Unit UK says

"colonies can be fully inspected on fine days when the bees are flying freely because this means that they are no longer in a cluster and therefore the colony is suitable for inspection. Colonies should not be inspected below 10°C, but they can be opened quickly between 10 and 14°C" (National Bee Unit, 2025).

I'm not aware of any recommendations suggesting that day length should be considered for spring inspections, but there does seem to be logic in the idea that stress may be caused by inspecting colonies that are still in a winter diapause state. If honey bees do indeed share a similar circadian rhythm to the wasp as discussed earlier, there may be an argument for using day length as a guide of the timing of spring/autumnal inspections. As a thought experiment, I decided to plot a graph (see figure 4.9.3) showing the day length by month for my apiary and then add to this graph the known day lengths that turn on and turn off a diapause state in wasps. For my own beekeeping practice I wondered what this would say about the timing of the first spring inspection. Using this day length model gives an inspection window of between 18th March and 29th August. I'm not convinced, but it has been an interesting thought process all the same and I have resolved to carry out my next spring inspection when the day length is above 14 hours and when the temperature is above 14 degrees Celsius - my own easy to remember 14 rule!

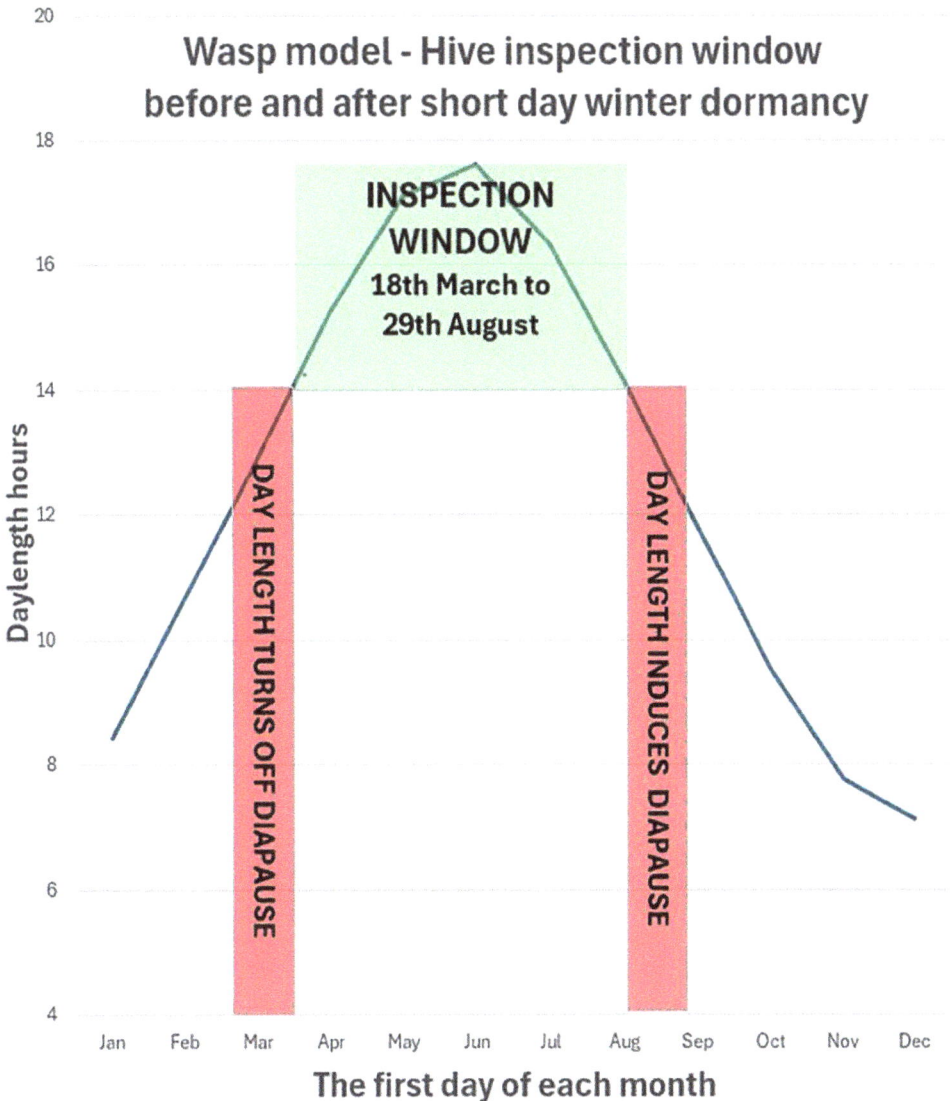

Figure 4.9.3 Wasp model for hive inspection

Accessing honey stores

Another similarity between the debris from March and September is a whitish deposition of sugar crystals, chewed wax and wax flakes away from the cluster (see figure 4.9.4 and 4.9.5). It is plausible that this debris is made by 'in-hive forager' bees who make quick forays from the warm core through the mantle to bring food back to the core of the cluster. The presence of fresh wax flakes also suggests that these same food collector bees are also moving wax flakes out from the core to do some kind of comb repair. These finds were surprising because I had thought that wax flakes were produced mainly in the spring and summertime. The reason why most of this food store debris is located at the furthest point from the cluster is uncertain. It will be interesting to see if this deposition continues in October.

Figure 4.9.4 Accessing food stores?

September varroa

Each year, I reluctantly treat my bees with a varroacide. I was planning to do the same this year, but because the visible mite population is so much lower than the NBU recommended level for treatment, I'm having second thoughts. Figure 4.9.6 shows the mites collected from the screen and figure 3.5 shows the varroa count for each month.

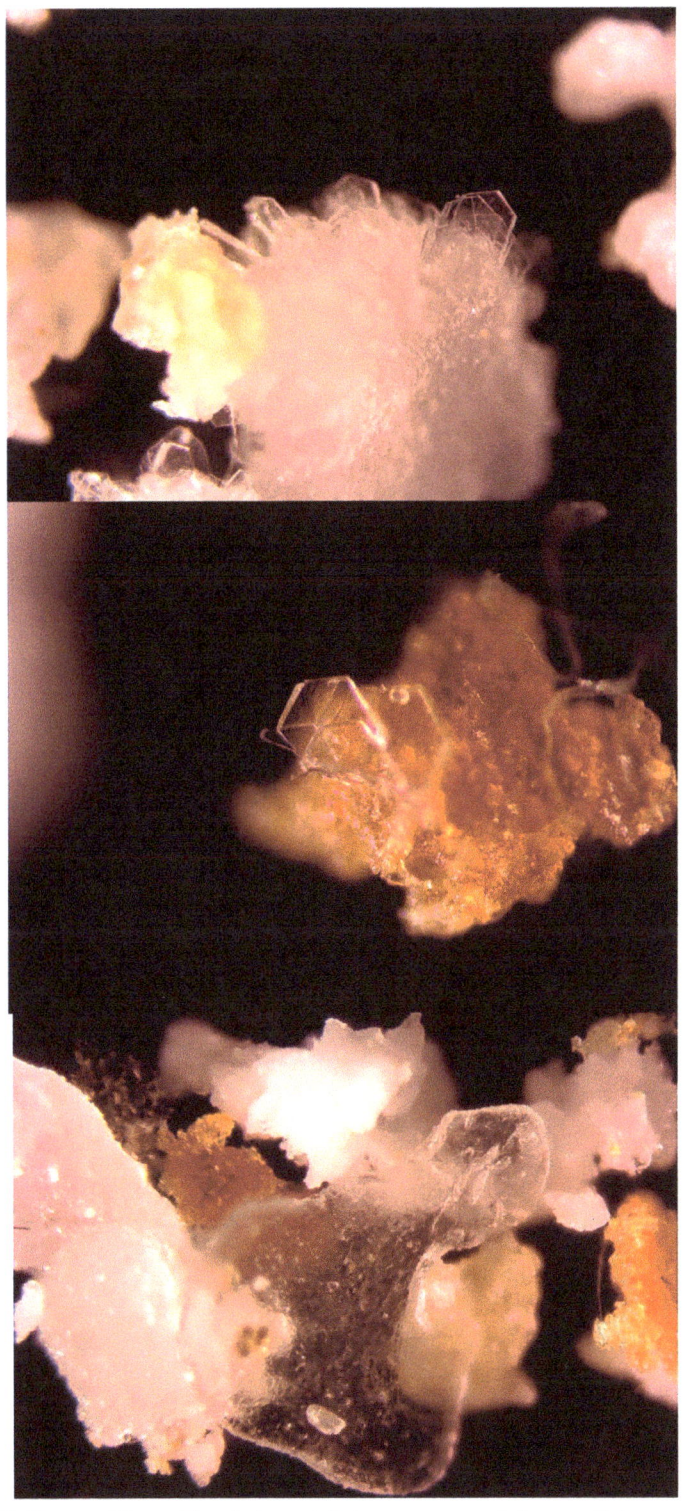

Figure 4.9.5 Composition of food store debris - sugar crystals, wax flakes and chewed wax

Figure 4.9.6 Varroa visible by eye

Figure 3.5 doesn't tell the whole story because microscopy reveals many more mites in the debris. These are mostly broken, damaged and invisible to the naked eye. The assumption is that this damage has been caused by a hygienic trait in the bees. Some simple maths shows a 7:1 ratio of damaged invisible mites to whole visible mites in the debris. It is surprising that this level of mite damage was only witnessed in September. I struggle to find an explanation for this. If I include these 'hidden' mites in the total count this hive meets the threshold for treatment, but if I exclude them it doesn't.

Figure 4.9.7 shows a few examples of damaged varroa, including single legs in the debris, exoskeletons that have had their soft inner parts scooped out, and legless varroa. What seemed like simple advice initially now seems a lot more complicated. Perhaps I should be reassured that the NBU guidance doesn't say get out a microscope and include those in the count! For now, the plan is to delay a decision and keep monitoring.

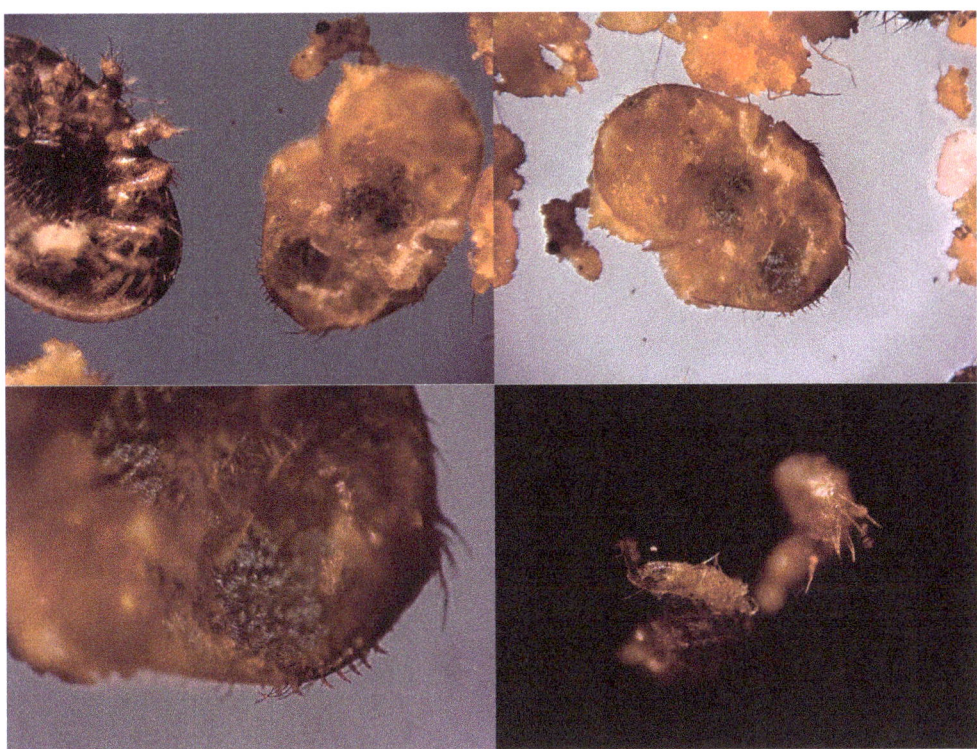

Figure 4.9.7 Varroa parts shown by microscopy

Figure 4.9.8 shows pumpkin shaped phacelia, spiky Asteraceae and rectangular Himalayan balsam pollen continue to show up in large numbers in the debris. Less common is the enormous pollen structure bottom left in the diagram. At 0.1mm in diameter it is either mallow or marrow of which we have both. It's confusing trying to tell the two apart. The key seems to be able to identify caps that sit on the pores. Marrow has caps attached and mallow doesn't. Confusingly the caps on the marrow can drop off and can look like pollen in their own right.

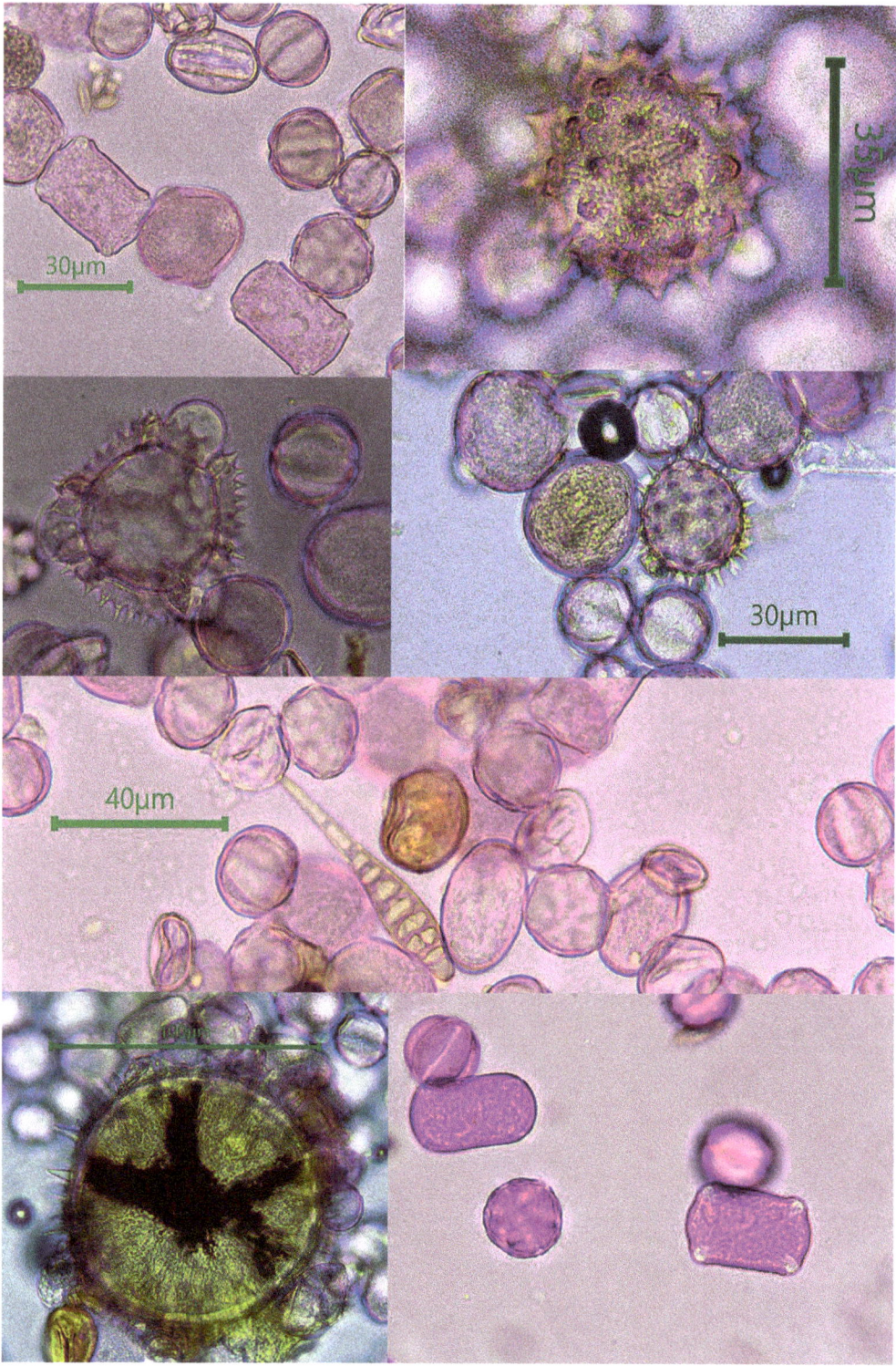

Figure 4.9.8 September pollen

4.10 October

Removing or adding stores

Honey bee debris collected between 1st and 28th October

Minimum temperature -1°C

Maximum temperature 17°C

Mean temperature 10°C

Figure 3.1 shows a surprising 100% increase in the October debris compared with September. Yet a comparison of brood debris on the inspection boards for these two months (figure 4.9.1 and 4.10.1) suggests that there has been a reduction in the volume of the cluster within the hive. Why would the debris weight increase when the cluster seems to be decreasing?

This increase in the weight of debris can be explained by debris created from bees accessing food stores. See the continued deposition of sugar crystals, wax flakes and chewed wax in the bottom left corner shown by the inspection board (figure 4.10.1). The heat maps show this pattern more clearly. The inspection board also showed similar debris in many areas away from the cluster, albeit in much smaller amounts. If the weight of this 'whitish' debris is subtracted from the total value, then the weight of the October debris would be almost half the value shown by figure 3.1.

As in September, this whitish debris (figure 4.10.2) had a similar composition, including three things: sugar crystals, virgin wax flakes and freshly chewed wax (4.10.3)

Figure 4.10.1 October debris

Figure 4.10.2 Whitish debris

Figure 4.10.3 Sugar crystals, virgin wax flakes and freshly chewed wax

The continued build-up of whitish debris suggests a different kind of activity away from the brood nest. Perhaps this shows a process of removing stores and a process of rebuilding and remodelling wax structures. Why else would the bees use precious energy bees to carry fresh wax flakes from the warmth of the cluster to this outer cold edge of the hive?

An alternative explanation could be a late autumn flow of ivy nectar, with bees using it to make stores and wax. In other hives it is not uncommon to find wax build up around the mouths of feeders left on late, and to observe brace comb glueing frames together that were moveable in September. I did look for evidence of pollen around this white debris thinking it may explain whether this was foraging activity but found none, so dismissed this idea. It's difficult to explain without watching the behaviour of the bees, but if I was asked to get off the metaphorical fence I would hold one finger in the air and say that I prefer the idea that this debris is from bees accessing their honey stores. I am choosing to believe that virgin wax flakes are a good sign at any time. They suggest that there is a nectar flow, brood rearing, the presence of a queen, the presence of pollen, a young workforce, and a suitable brood nest temperature.

Varroa

Figure 4.10.4 Whole varroa

In September, there were many wounded and limbless mites. Last month's debris was more akin to a crime scene, and it was difficult to find a whole mite. This month all of the 34 varroa found in the debris were complete. No damaged varroa or body parts were visible using microscopy. Figure 4.10.4 shows a typical example of the 35 mites found on the inspection board. I am struggling to explain why in one month there will be lots of damaged mites and in the next none.

An average daily drop rate of 1.1 mites per day has remained constant for two months, suggesting that the mite population is not increasing. This is still well below the NBU recommended value for treatment. If the mite drop started to increase, I intended to treat the colony this month, but I will continue to monitor the varroa population and look for evidence of damaged mites for now.

Early in the year, I had real concerns about high chalkbrood levels and the impact this may have had on colony health. I am pleased that in October there are almost no chalkbrood structures in the debris. Figure 3.4 shows that signs of chalkbrood in the debris are at their lowest this year. The bees have managed this infection for themselves. On the inspection board, It no longer looks like an over generous sprinkling of black pepper. This development is important learning for me, because what seemed like a high infection rate was not catastrophic for the colony and the bees seem to have managed to resolve this issue themselves. In May, I was considering replacing the queen, thinking that this would help to reduce the chalkbrood load but with hindsight this was not necessary. It's this kind of dilemma that can give a beekeeper sleepless nights. The best course of action is not always obvious. A never-ending cycle of learning for the beekeeper through the seasons is the only way we can improve our chances of success when faced with tricky decisions like this. I wonder if the brood break reported in June could be a contributory factor to this reduction in chalkbrood?

Bee hair deposition continues to be a fascinating topic. In October there has been a reduction in the number of hairs found in the debris. Figure 3.3 shows an estimated 760 hairs per gram of debris compared with 2730 hairs per gram of debris for September. I am intrigued about the idea of identifying these body hairs. A fascinating study has made this possible by creating an open-source gallery showing the morphological structure and distribution of hairs on different parts of the bee body (Kahn & Liu, 2022). Using their gallery of hair images, it should be possible to compare the hairs found in the debris and identify the body part from which it came. It would be a mammoth task but could be enlightening to know the origin of the hairs being found in the debris.

Alternaria in the debris

Over the last year, different species of fungus from the genus alternaria have been found. Some of these carry out ecological services like plant decay and some are parasites.

In October there was a new find which I believe to be *Alternaria solani* (see figure 4.10.5). This species is known to cause blight in tomatoes and potatoes. These fungal structures are very small, some being as small as a tiny pollen grain, and these will be easily carried by the wind or caught by a bee in flight. When they meet leaves, each segment of the fungi grows a tube. These tubes are believed to grow through holes (stomata) on the underside of the leaf. From these tubes then grow mycelium which then cause yellow brown rings on the leaf and reduce photosynthesis causing leaf decay. It's feasible to suggest these are accidentally trapped in bee hairs during flight.

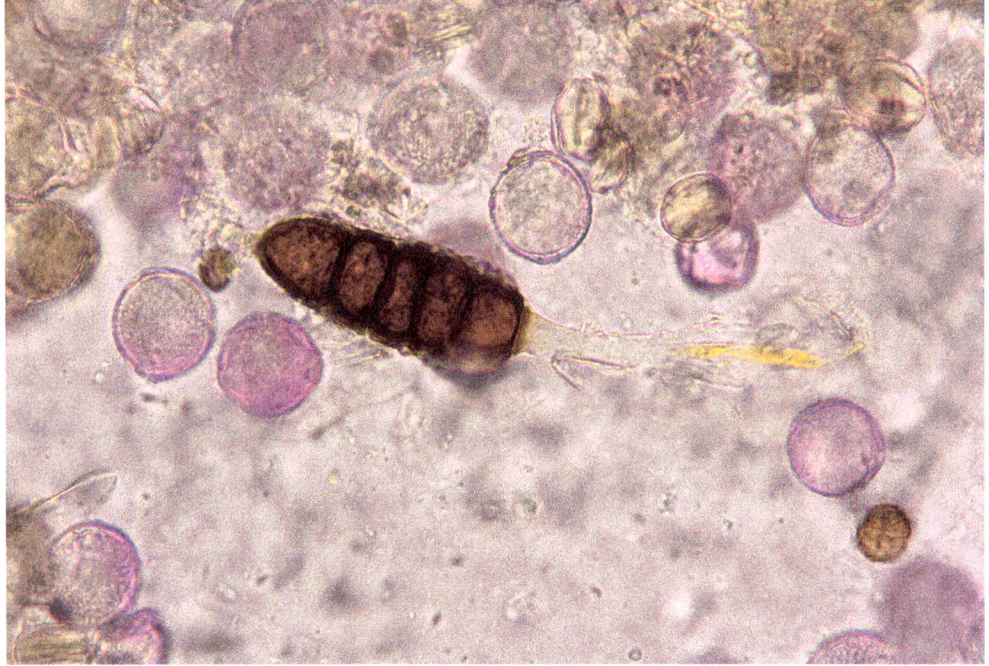

Figure 4.10.5 Alternaria sp

Stellate trichomes

In the October debris there are hundreds of trichomes (see figure 4.10.6). These were also a common find in the preparatory work for this study back in the Autumn of 2023. In both years these trichomes were commonly found amongst ivy pollen, so it seems logical to assume that they are associated with ivy. However, field verification would be needed to be more confident about this conclusion. These cellular structures are so common amongst the debris that I wonder if these are also stored in bee bread or eaten by the bees. It seems unusual though, because a major function of trichomes is thought to be a plant defence strategy.

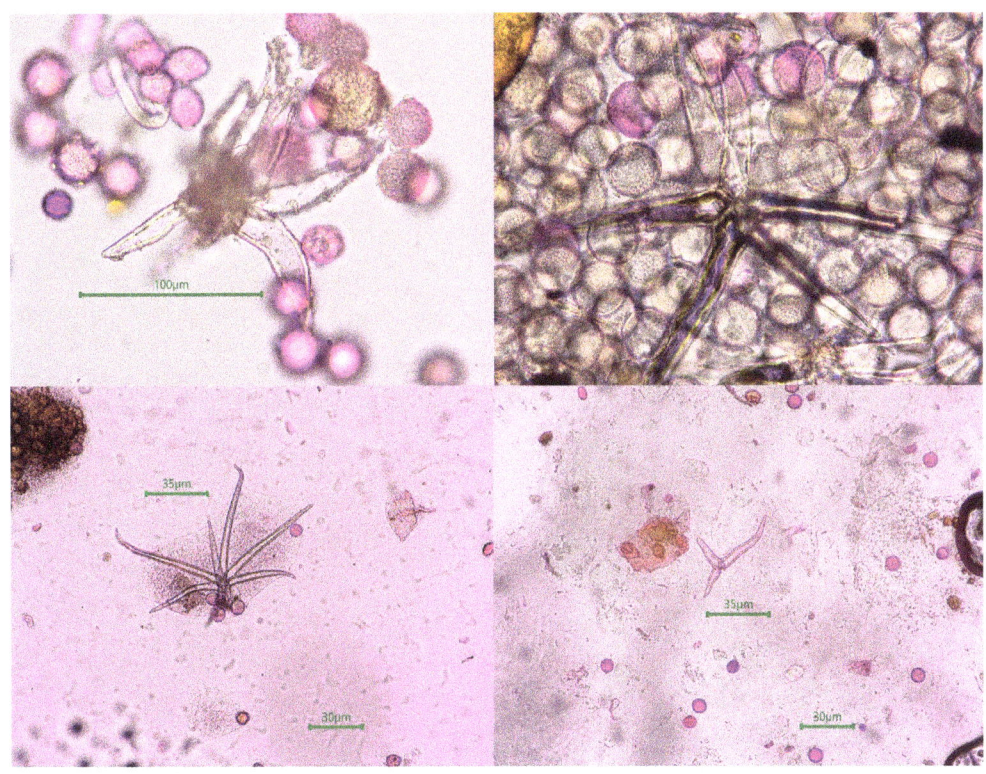

Figure 4.10.6 Trichomes surrounded by Ivy pollen

Pollen in the October Debris

Starting top left, figure 4.10.7 shows lots of ivy (bottom right), spiky Asteraceae pollen and an unusual 10 pore structures around 35um that resemble the teeth on a circular cutting blade. This circular saw pollen is borage. It is similar in appearance to comfrey, to which it is related. We planted a few borage plants about ten years ago and have never needed to plant any more. It is very effective at reseeding and looking after itself. We also have lots of comfrey. I have learned that it is possible to differentiate between comfrey and borage pollen by counting the pores - comfrey only has nine pores. The bottom left pollen is honeysuckle, of which we have many late flowering varieties in the garden.

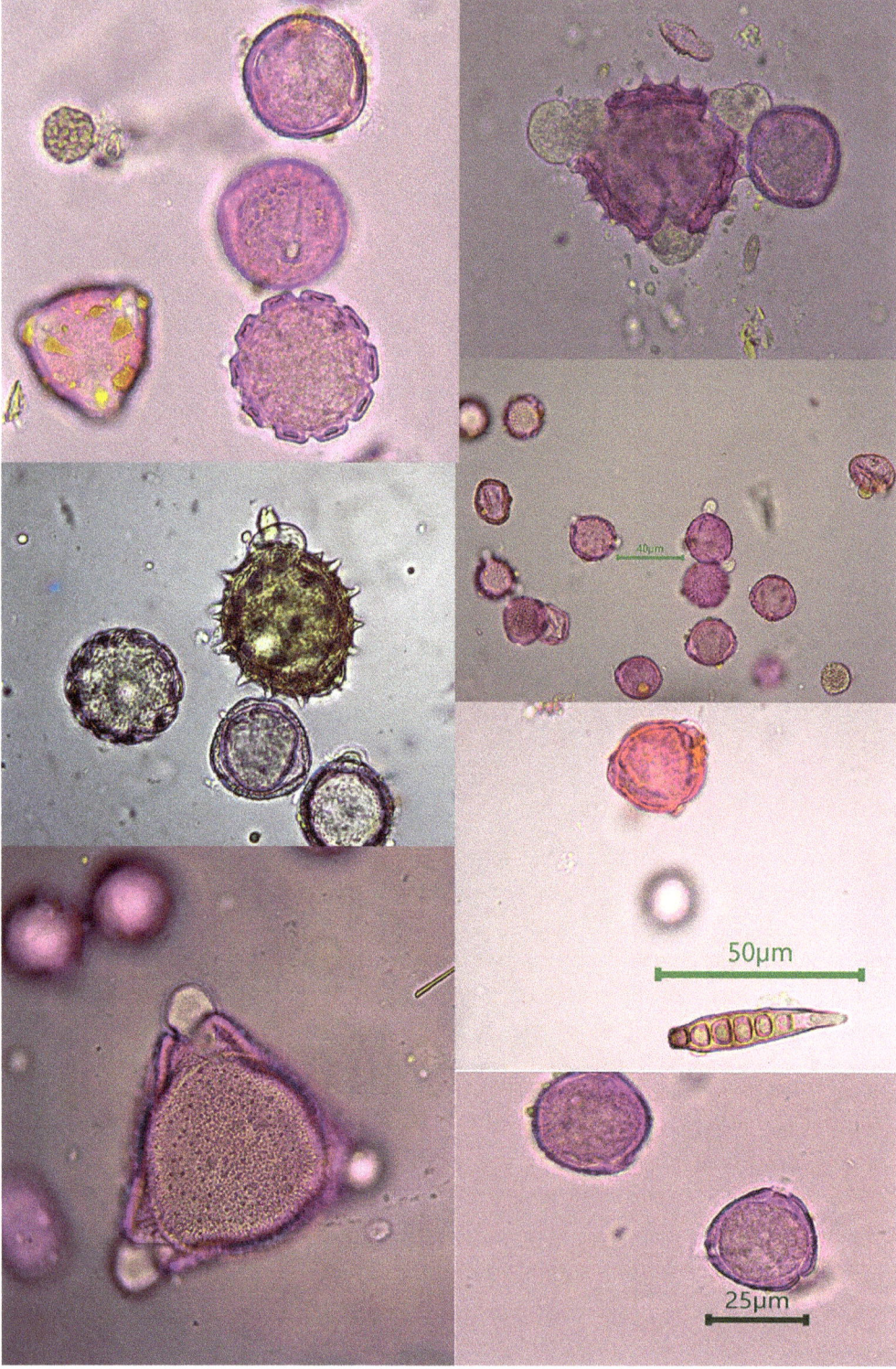

Figure 4.10.7 October pollen

4.11 November

More fungi

Honey bee debris collected between 1st and 28th November

Minimum temperature -6°C

Maximum temperature 16°C

Mean temperature 6°C

The rate of debris deposition in November is similar to January and February of the same year. Figure 3.1 shows that the deposition is just over 1 gram each week for each of these months. The debris looks finer, although I have not measured particle size by month so cannot state this as a fact. There is no longer the same build-up of 'food-store' debris in the bottom left corner as found in the previous two months.

This colony must be working hard to maintain critical values that are not too high or not too low, but just right for survival. The debris pattern (figure 4.11.1) is an outcome of this activity. It is remarkable that the cluster will live in a similar position in the hive for several months, maintaining homeostatic control by regulating its population and controlling the movement of resources and waste materials in and out of the cluster. This form of dormancy includes behavioural and physiological responses that govern the population size, maintain a constant temperature and many metabolic processes. Part of this is the remarkable ability of forager bees to break from the warmth and safety of the cluster, to take advantage of any external temperature rise and access winter pollen, water and nectar.

Figure 4.11.1 November debris

It is untrue that bees only leave the hive on warm days. In one study Scottish honey bees were commonly observed collecting water at only 4 degrees Celsius (Chilcott & Seeley, 2018). Figure 4.11.2 shows the temperature to be barely above zero on sampling day.

Figure 4.11.2 The temperature on sampling day

The cluster needs a supply of water, even on cold days, for their metabolism and for brood development. Leaving the hive is risky business at any time, but especially so when it's cold. Foragers need a thoracic temperature of above 25 degrees Celsius just to power their wing muscles for flight. Exposure below this value runs the serious risk of cold shock and death. The conundrum is how bees manage to maintain their body temperature to collect cold water and return home. Chilcott and Seeley describe research showing that when a bee is at the water source, its wing muscles are activated (shivering) to keep a thorax temperature above 35 degrees Celsius. We are left wondering if the same thing happens when bees collect cold winter pollen and what happens to these 'cold' foragers when they return to the nest. Are they allowed to freely enter the core, or will they need to be warmed up a little before being allowed back in? It seems likely that honey bees break from the cluster in cold weather to perform a variety of other essential hive maintenance tasks, such as feeding, waste removal and recycling.

I have learned to be more confident that this debris shown by figure 4.11.1 suggests that the bees are overwintering just fine, firmly in a diapause state and it seems likely that the debris pattern will remain similar until March next year. This view is further supported by the debris patterns from additional colonies shown by figure 4.11.3. These photographs were shared by a beekeeper to highlight a 'healthy' debris pattern from an apiary in Slovenia.

The nutrition required by the colony to carry it through the winter varies according to the strain of the bee and colony size. The NBU UK says that 10 Kg of honey is needed for honey bees to get through the winter and recommends feeding to supplement any shortfall. That's five full brood frames of honey. This colony has double that, so hopefully no additional feeding will be required. I will continue to monitor and will have no hesitation in feeding fondant or sugar syrup if there is any risk of starvation, or as a temporary intervention if a colony needs a boost.

Some beekeepers take most of the honey stores and feed sugar syrup claiming that it does no harm. In a club or group situation, I tend to be polite, not saying what I think of this strategy, not wanting to offend and respecting each beekeeper's right to make their own decisions. However, as part of the process of writing this book, I will get off the fence and say what I really think!

There are as many ways of beekeeping as there are farming and fishing. It may be unpopular to say, but I am pleased that the cost of artificial inputs, like agrochemicals and granulated sugar are increasing because it is changing the economic argument for their use. Farmers, like beekeepers, are looking for more financially and environmentally sustainable approaches, some because they want to work in different ways and others who are being pushed by cost. In my opinion, the beekeeping community has become too reliant on cheap sugar as a 'magic' fix. In some cases, sugar syrup has become a nutritional sticking plaster for a landscape that has a deficiency of floral nectar and a beekeeping approach that looks for quick 'profit' rather than a more sustainable way of doing things.

Figure 4.11.3 Overwintering just fine (Kind permission from Miha Meteiko, Slovenia)

Through writing this study I have learned how a combination of honey yield and debris can be proxy indicators - a valid and valuable local record of the colony and different environmental conditions. Like tree rings, corals and ocean sediments, it is my view that there is similar worth in studying debris to better understand bees and the local environment. I feel so much better informed about the biophysical characteristics of this one hive in its external environment. As one small illustration and because of this study, I have an improved awareness about the forage used by my bees. For example, field bean pollen has been found in most months' debris, and it has a high nutritional value for the bees, it contributes to high honey yields and I can now predict when it will be in flower. While out for a November walk, I made the obvious connection with the more than a hundred acres of field bean seedlings growing right next to my apiary (see figure 4.11.4). It's been sown as part of a drive to improve depleted soils but will also have a positive impact for local pollinators. Now's the time to make predictions about the likely flowering dates for these and for other local flowers and make beekeeping plans for the coming year. November is a very good time to start making these plans!

Figure 4.11.4 Field bean seedlings with our smallholding tree boundary in the background

More fungi in November

It seems that the methods to classify the fungal genus Alternaria have been debated for many years resulting in disagreements, differing approaches and revisions. One approach has been to classify the structures based upon their morphological characteristics. This has led to difficulties because there can be an overlap of characteristics, and these can be affected by environmental conditions and intrinsic factors. The most comprehensive guide to classification based upon morphology has been produced by Simmons who studied Alternaria for over 50 years (Simmons, 2007), producing many articles and papers on the subject. This work, along with recent developments in molecular biology and a DNA based approach, has led to a total of nine Alternaria species being discovered. To accurately identify this kind of fungi you need to consider the size, shape, structure along with an analysis of the DNA.

Using only pictures from a summary paper (Lawrence, et al., 2016), my best identification effort suggests that the November debris contains three different species. Figure 4.11.5 shows these from left to right

1. Alternaria japonica which causes decay in brassicas,

2. A.infectoria responsible for causing opportunistic infections

3. A.scirpicola which are generally single chains containing several units.

Whether or not these best guesses are correct seems less important than acknowledging what appears to be an increased diversity of Alternaria species at my home apiary in the autumn.

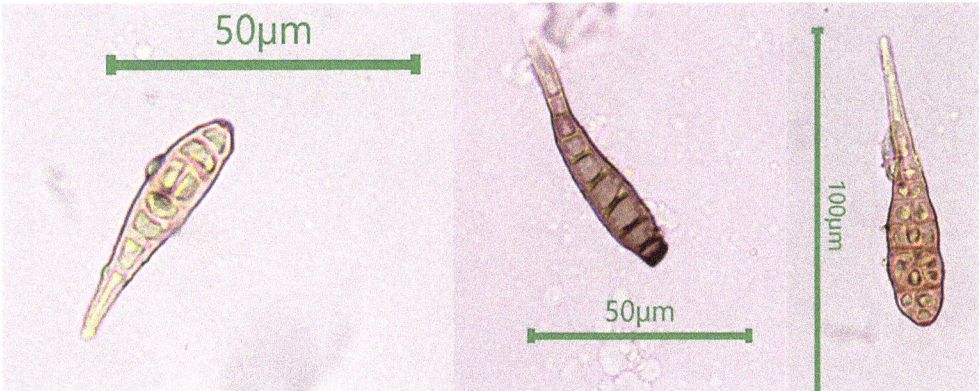

Figure 4.11.5 November Alternaria

Chalkbrood and bee hairs

It is still possible to find the reproductive structures of chalkbrood in the debris by using the microscope. Figure 3.4 shows that in November there were an estimated 2,400 chalkbrood cysts in the debris. This is by far the lowest level recorded. Figure 3.4 shows a continual decrease in the estimated number of these structures since May. While it is reassuring to see this reduction, the continued presence of this fungus suggests that it has the potential to take hold again when the conditions are right. It's one to continue watching and learning from the changes in numbers over the months.

The estimated total number of bee hairs in the debris has risen from almost 10,000 to 24,000 in the last two months, compared with an estimated 74,000 hairs in August. As discussed in earlier months, it may be helpful to eventually identify these hairs to explain their presence in the debris.

November pollen in the debris

Of the eleven months sampled so far, November contains the greatest diversity of pollen for any month in 2024 (figure 4.11.6). Many of these species have not been in flower for months. It is feasible to say that this debris is from honey bees accessing preserved food stores, moving and consuming bee bread within the cluster. It is good beekeeping knowledge to know that your colony has a balanced diet throughout the winter. There are many old favourites from left to right: field beans, rosebay willow herb, an unknown 5 pore structure which may be field pansy, ivy, phacelia, Asteraceae, rapeseed, many stellate trichomes, borage, heather and dandelion as described in previous months.

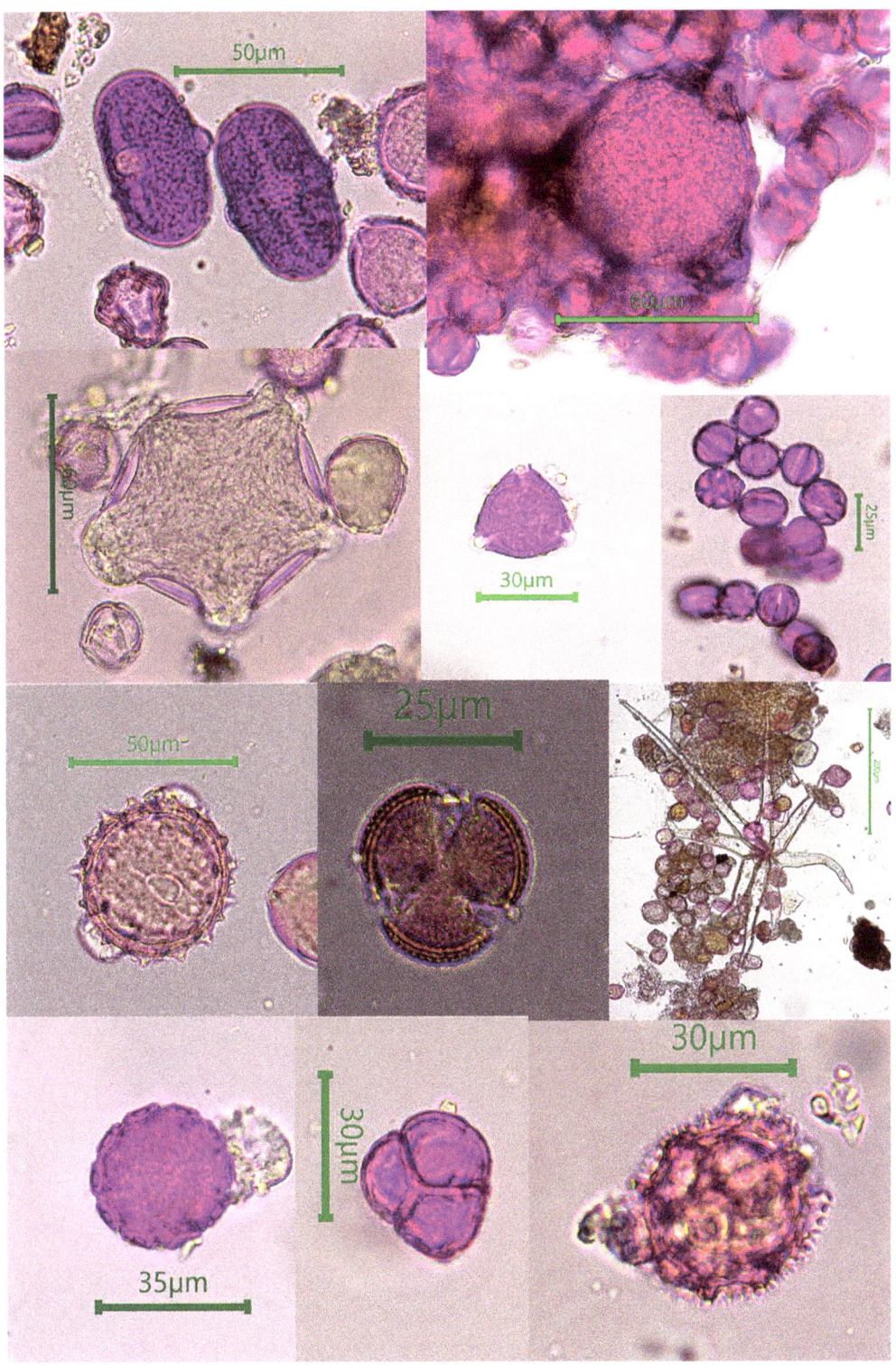

Figure 4.11.6 November pollen

4.12 December

Return of the mites

Honey bee debris collected between 1st and 28th December

Minimum temperature -5°C

Maximum temperature 14°C

Mean temperature 7°C

Figure 3.1 shows 4.03g of debris deposited on the inspection board in December, compared with 4.08g collected the previous month. The values are similar to deposition rates for January and February. This represents a constant rate of deposition of around 1 gram of debris each week for a third of the year, which says much about colony activity. Figure 4.12.1 compared with previous debris shows a similar distribution for these months. The heat map shown by Figure 4.12.2 shows that there has been a shift in brood activity towards the eastern side of the colony.

The inspection board shows a similar debris pattern as recorded in the previous few months. A mixture of wax flakes, sugar crystals and chewed wax can be seen without a microscope, situated away from the cluster. I am now more confident that the debris shown in detail by figure 4.12.3 is from bees accessing their stores as discussed in previous months. I am less confident in explaining the pool of liquid on the inspection board (see quadrant C2 on Figure 4.12.1). Back in January a similar pool was found, and when I tasted the liquid, noted its sweetness and resolved to test the solution if it happened again. Figure 4.12.4 shows this pool of liquid being tested to find a concentrated sugar solution.

Figure 4.12.1 December debris

Debris heat map November (g)

1	2	3	4	5	6	7	8	9
0.05	0.07	0.21	0.48	0.55	0.51	0.26	0.104	0.09
0.05	0.09	0.19	0.3	0.3	0.28	0.14	0.03	0.05
0.01	0.01	0.01	0.02	0.03	0.02	0	0.01	0.01
0.03	0.01	0.01	0.01	0.01	0.01	0.01	0.01	0.01
0.14	0.18	0.42	0.81	0.93	0.82	0.41	0.21	0.16

Total 4.08

Debris heat map December (g)

1	2	3	4	5	6	7	8	9
0.08	0.08	0.08	0.34	0.17	0.37	0.25	0.14	0.036
0.06	0.07	0.1	0.21	0.32	0.25	0.14	0.05	0.03
0.06	0.06	0.06	0.06	0.06	0.06	0.06	0.06	0.06
0.05	0.05	0.05	0.05	0.05	0.05	0.07	0.08	0.06
0.25	0.26	0.29	0.66	0.8	0.73	0.52	0.33	0.19

Total 4.03

4.12.2 Heat map for November and December

Figure 4.12.3 Debris away from the core

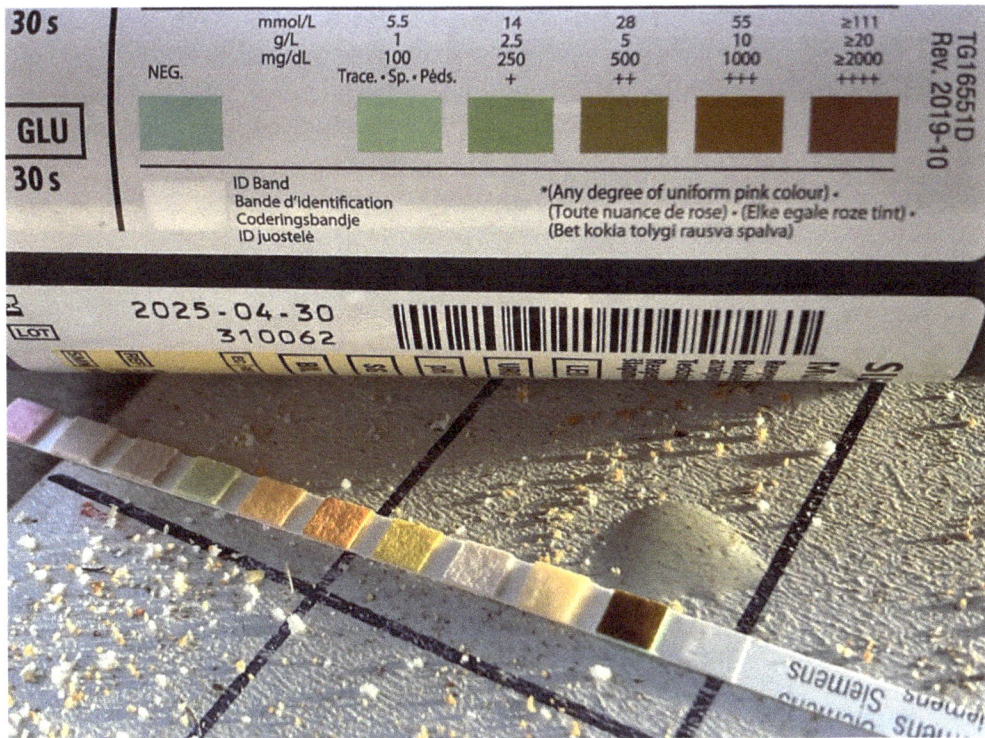

Figure 4.12.4 A concentrated sugar solution

December debris shows a healthy population of reproducing mites. Figure 4.12.5 shows a small sample of these mites. They look identical to the ones recorded in February and makes me wonder if a population of Carpoglyphus lactis has been living amongst the bees throughout the year. For some reason the conditions now are more optimal for their reproduction and/or there are fewer honey bees available for biocontrol.

As discussed in February, it seems plausible that the mite population dynamics are related to hive humidity. One study states that "honey bees are highly vulnerable to these mites which can rapidly infest and consume [stores]...leading to a weakened colony and potential colony collapse" (Nguyen, et al., 2024). This study found that the detection of mites was more than four times greater in overwintering colonies compared with summer ones.

To better understand these mites at my home apiary, I placed three brood frames that had a mixture of open and capped honey into an empty box with no bees. I sealed the box and placed an inspection tray underneath to collect any debris. One month later, the inspection tray showed a fine golden-brown powder on the inspection board (see figure 4.12.6). On closer examination this mite debris could be seen to be moving due to what looked like 100,000s of mites living in this debris. I removed the frames and could see the stores that had been consumed by these mites. It became clear that these mites are highly efficient at exploiting resources within the hive. They have the potential to create a prolonged pressure on food stores and become a potential health hazard. While it seems that healthy bees are very efficient at controlling these mite populations, they are yet another potential stress factor on honey bees, and when combined with other stressors, can have catastrophic consequences for the colony. In future years, I will certainly be paying more attention to these mites.

Figure 4.12.5 Mites

Figure 4.12.6 Mite debris

Figure 3.3 shows that there are over an estimated 10,000 body hairs per gram deposited in December. This is more than double the count of hairs per gram recorded in any previous month. Figure 4.12.7 shows a few examples of these. Some are long and appear to be whole and pulled out at the root, whilst others are very small and appear broken/fragmented.

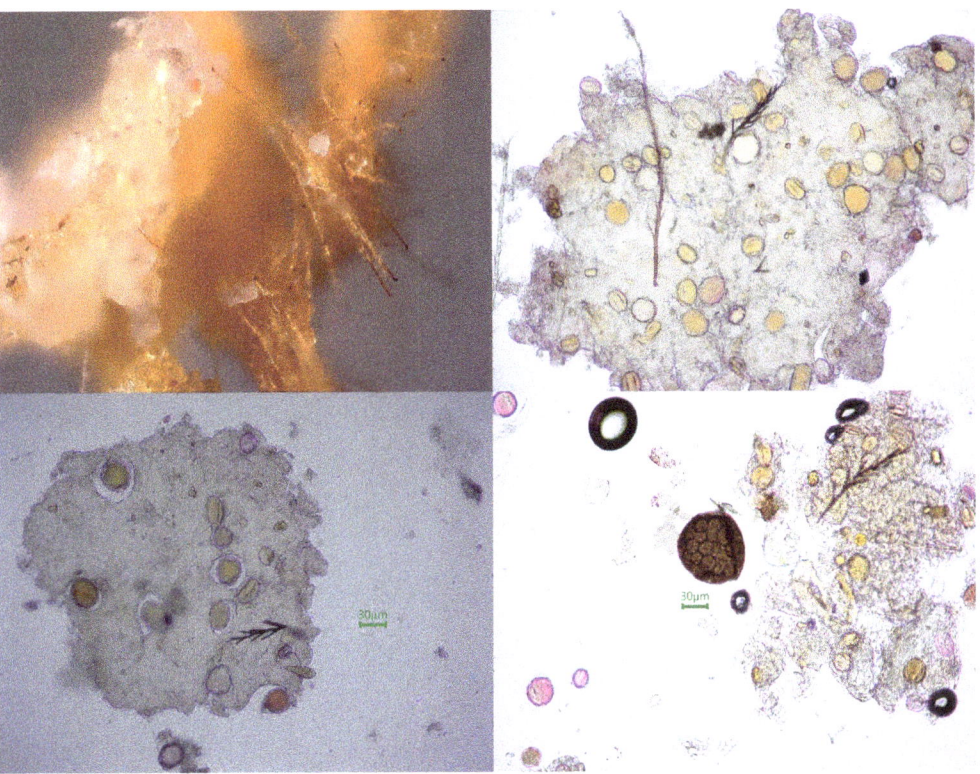

4.12.7 Bee body hairs

Figure 3.4 shows that while the number of chalkbrood structures per gram of debris still seems quite low, there has been a four-times increase in their deposition compared with the previous month. Throughout this study I have been surprised by how much variation there is in the number of these structures and just how quickly the situation can change.

Figure 3.5 shows that the varroa drop on the inspection board in December is similar to the drop rates for September, October and November. Still below the NBU level recommended for treatment. No dismembered or damaged mites were found in the debris as discussed in previous months. As with the chalkbrood situation, I am mindful just how quickly this situation can change and of the importance of monitoring.

Figure 4.12.8 shows the main pollen deposited in the December debris. The top two pictures show field bean pollen. The top right of these is I believe preserved bee bread with field bean pollen. This pollen has become a favourite, showing up almost every month this year. This may have been accidently dropped whilst feeding. Also, ling heather, phacelia, Asteraceae, Himalayan balsam, a large mallow or marrow and an unknown 3 pore structure of about 20um.

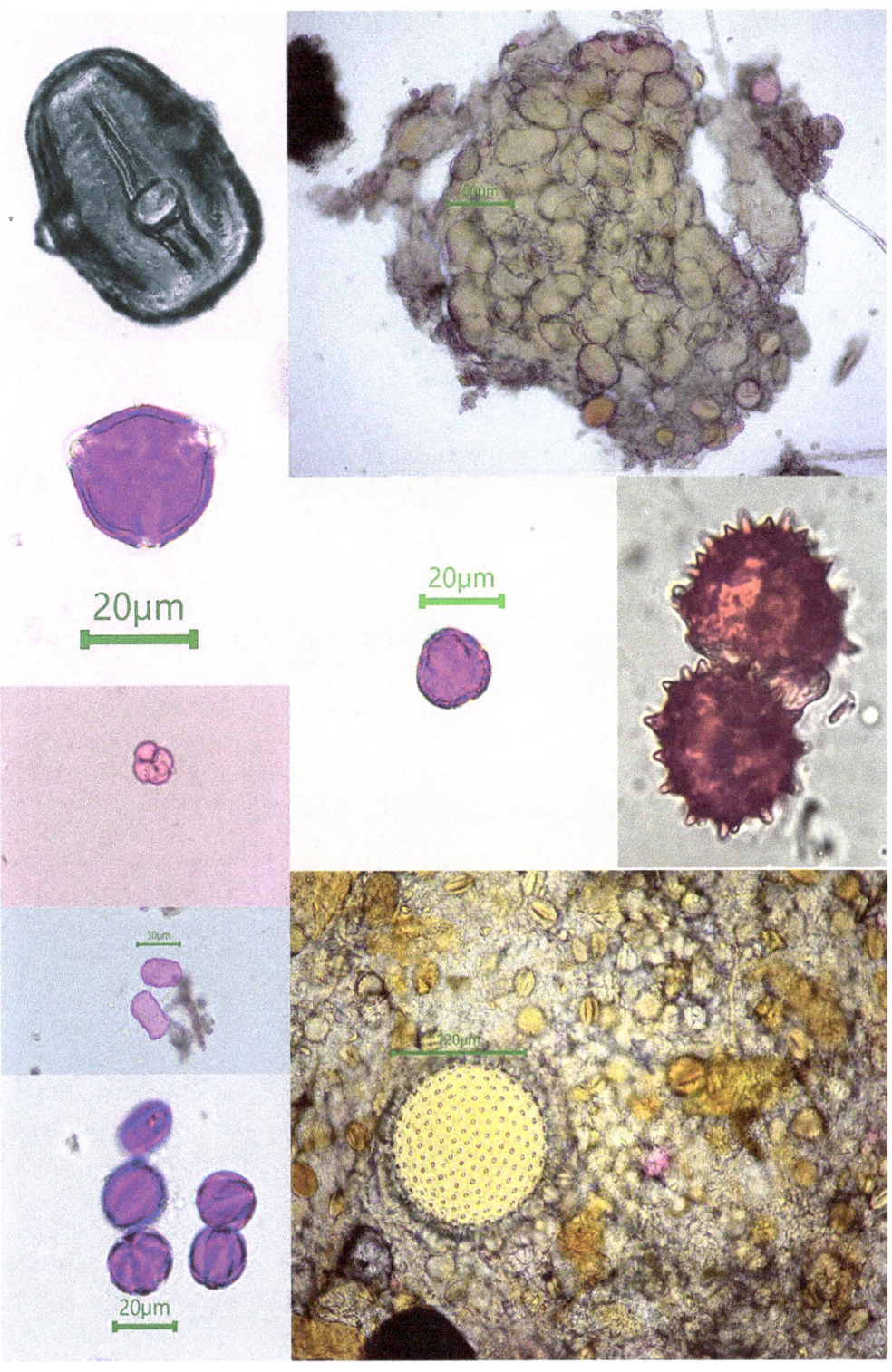

Figure 4.12.8 December pollen

5 Conclusion

Next time you flush the loo, visit the bottle bank, separate and wash packaging or compost materials give a thought to the honey bee hive biome. This biological community of honey bees, fungi, mites and bacteria are also busy turning the 'waste' products of colony activity into materials that can be reused in the hive or into usable nutrients that can be used by bees or other organisms. Most of the 'waste' materials that sustain this biome have been manufactured from biomass collected by honey bees - the sugars, wax, pollens and resins.

Microscopy has also showed many man-made fibres in the debris, demonstrating how honey bee foraging provides a mechanism for bees to inadvertently act as bio-samplers by collecting fine human made particles and that these most likely end up in the debris deposited on the inspection board. The idea that honey bees can act as pollution detectors is not new and was well described by Steen (2016) in his paper 'Beehold, the colony of the honeybee as a bio-sampler for pollutants and plant pathogens'. There are numerous studies recording agrochemicals and other pollutants in hives. Steen states that "bio-indication with honeybee colonies has more applications than have been exploited so far. Further research can make a change".

It seems plausible to suggest that honey bees, or rather the debris produced by them, are the 'canaries in the coalmine', acting as potential indicators of the health of the colony, the ecosystem and potential danger. My bees forage in areas that have been treated with agrochemicals, so it seems highly likely that pesticides, fungicides and herbicides will be found at different concentrations in the debris at different times. Certainly, it's an interesting research question to explore further. As honey bees forage over such a large area, could it also be that there is a 'bioaccumulation' effect of human-made particles found in hive debris compared with the wider environment? Using the debris for this kind of research would also seem to have the advantage of being less invasive compared to sacrificial sampling methods which involve killing bees and colonies to answer research questions. In addition, because collecting and analysing debris is relatively cheap and simple to do, it becomes more accessible to citizen scientists who are interested in honey bee research. I'm aware that I've only skimmed the surface of what can be learned by studying honey bee debris. For this reason, I have been freezing monthly samples of debris and hoping in the future to use these for some kind of diagnostic testing. Perhaps you will also have questions or research ideas to study using a timeseries of honey bee debris. I know I do.

In the introduction, I started to make a case for 'Bottom up Beekeeping' – a less invasive way to understand the colony without disturbing the brood nest. Honey bee debris has taught me a great deal about honey bees and the hive biome. I hope that there has been learning for you too. The monthly descriptions have shown

- the debris studied in this project could only have been produced by a colony with a laying queen.
- A queen that went on to increase her rate of egg laying to produce a colony that would produce enough honey so that no supplemental feeding was needed, while also producing a healthy surplus of honey for personal consumption.
- A colony that also managed a significant chalkbrood burden and controlled a varroa population with minimal human intervention.
- A monthly 'audit' of pollen in the debris showed the pollen diet consumed by these bees

All vital information for the beekeeper.

Perhaps you have resolved to look closer at the debris produced by your bees – that would be wonderful. Over the months. my learning journey also manifested itself in a deep critical reflection about my own practice and relationship with bees. This was an unexpected but very welcome part of the research process.

My relationship with bees has formed over nearly fifteen years, from trial and error, making mistakes, from being the teacher and the learner. No doubt I have learned lots about bees, but with hindsight I have learned much more about hubris and a need for humility in how humans interact and care for bees and the environment. In beekeeping, this hubris is an excessive belief in humankind's knowledge about bees that so often displays itself in the practice of trying to outsmart the environment and the honey bee.

For more than a decade I have looked down on the frames in my hives, paying scant regard to what was happening below or what went before. Today, a 'top-down' top opening hive has become the industry standard on planet earth. This was recently illustrated to me by friends who cycled by tandem from Canada to South America. They sent me pictures of hives spotted along the route. Every time it was the same beehive design. Had Langstroth been successful with patenting his hive invention, his descendants would be very wealthy indeed. Langstroth's relatively recently acquired knowledge of bee space transformed beekeeping at roughly the same time as society made the transition from horse ownership to fossil fuel powered transport. Today it is difficult to imagine travelling without cars, buses, trains and planes. Yet the reality is that people travelled the whole world, and also successfully managed bees before these 'great' inventions. It is interesting to note that hive design and transport developed at roughly the same time in a period of modernisation and industrial development.

There is little doubt that global travel and beekeeping are much easier now. The fact that we can take out and shuffle frames in any order, at will, creates an illusion that the beekeeper has total mastery of the colony. Sometimes I have wondered if this is more akin to Russian roulette, as we make decisions to change the brood nest structure, split the colony, catch the queen, clip the queen to prevent swarming, drone culling, restricting the cell size and much more. This mastery illusion can be illustrated by the variety of methods and opinions in the beekeeping community about the best way to proceed with each of these tasks. I include myself here and wish that I had a pound for every time that I have arrived at a hive, puffed smoke in the entrance, removed the roof, started removing brood frames in less time than it takes to boil a kettle. The study of bee debris is changing how I approach beekeeping.

As a beginner. I would inspect every frame, but slowly as I gained more experience, I learned how to inspect colonies in ways that reduced the impact upon the brood nest, and my bees became healthier and more productive for it. Developing the knowledge and skills to read the debris feels like a logical progression of this personal learning journey. A study of the inspection screen is no longer an occasional 'add on' to check varroa. It has become essential, and the information from that inspection determines how I proceed with my beekeeping plan. This bottom-up approach doesn't replace top-down inspections, it works hand-in-hand to improve the health and wellbeing of the colony and is part of the preparation to make positive hive interventions.

A recurring topic in beekeeping literature is the cumulative effect of stress factors on the colony. As I started to see these individual stress factors in the debris, I wondered about the stress factors that I couldn't see. Stress factors that I could see included varroa, wax moth, chalkbrood, mites, poor weather, poor forage, microplastics and various fungal species. Stress factors such as agrochemicals in the debris were not measured, but it seems highly likely that these will also be present at different concentrations at different times. Add to these stress factors a long list of human interactions that may cause colony stress, and it is easy to see how an individual stressor may well be sublethal but the cumulative effect of stressors has the potential to become deadly for bees. Table 5.1 shows examples of human interventions that may add to the cumulative stressors on the colony.

Table 5.1 - Cumulative stressors on the colony adapted from (Palgrave, 2022)

A colony that is/has
▸ Handled carelessly or opened when too cold.   ▸ Kept in an inappropriate, poorly insulated or dilapidated hive.   ▸ In a location which is too hot, too cold or too damp.   ▸ Is hungry or with inappropriate/insufficient stores.   ▸ Is split when too small.   ▸ Suddenly has its honey stores removed and/or is given a large empty space (e.g., honey supers or an empty brood box).   ▸ Stimulated early in the season and prematurely switches its nurse bees to foragers; this may leave the brood nest short staffed and particularly vulnerable to cold snaps.   ▸ Selected to collect less propolis will be more vulnerable to disease.   ▸ Practice that prevents the colony from building a propolis envelope   ▸ Diseased bees or that is infested with varroa and associated viruses.   ▸ Subjected to frequent, slow inspections.   ▸ Smoked too heavily, or with smoke that is too hot or acrid.   ▸ Subjected to manipulations which involve major disruption and/or cooling of the brood nest (e.g., splits, brood spreading, chequerboarding).   ▸ Transported between sites.   ▸ Suddenly being made queenless or becoming hopelessly queenless.   ▸ Exposure to noxious substances inside or outside the hive.   ▸ Is genetically/phenotypically inappropriate for a particular area.   ▸ is prevented from raising the number of drones it would like (worker foundation, drone culling).   ▸ Which repeatedly has its swarming urge/instinct frustrated.

Added together, these non-human and human stressors mean that bees must work harder and use more metabolic energy to prevent a negative effect on the colony. It is easy to think of health and disease in a binary way. Do you have measles, yes or no? Do your bees have dysentery, yes or no? And it is true, the presence of a disease is sometimes binary. However, health is often much more complex to define and represents different points on a continuum.

"We all know that when we get tired and 'run down' we are more likely to pick up the bug that has been circulating, or we may suffer from mouth ulcers, spots, thrush and various other infections. Those with chronic conditions like eczema, psoriasis and inflammatory bowel disease are also more likely to experience a flare-up...Stress has a profound impact on our endocrine system (hormones), it changes our metabolism, inhibits our immune system and alters our emotional state; there is also increasing evidence that changes to our microbiome (bacteria living in our gut) play an important role." (Palgrave 2022)

It seems like sensible and good advice to say that beekeeping practice should try to identify and reduce the number and level of stress factors on the colony, to approach the colony with more humility rather hubris. It is a similar approach to that of elite athletes who pay attention to every miniscule detail of their performance, to make small gains, knowing that the sum of those gains can create a world breaking performance. It seems logical that a beekeeper who pays similar attention to reduce stressors is more likely create a healthier and more productive colony. Understanding how to read the debris produced by a colony can be a way to understand the interaction of these stressors and to guide decision-making and preparation for making positive hive interventions.

As I write this conclusion in late February, I am mindful of all the 'good' advice about the timing of the first inspection. Some say after the spring equinox, others when the temperatures rise above 10°C, following flowering redcurrant and so on. I have tried them all and may even have added to this confusion by suggesting that day length may be an important consideration. This year is different, the debris suggests that this hive is overwintering just fine and for now I will leave the smoker on the shelf and continue studying the debris to make decisions about when and how to intervene in the colony.

Creating a timeline series of debris photographs often gives more information than a 'normal' open the lid style kind of inspection, a bold assertion which is supported by figures 5.1 and 5.2. These images, taken in less time than it takes to zip up a bee suit, show debris from the same colony for January and February for two consecutive years. Months in the year when colony activity and debris are at their lowest. What can we conclude from this timeseries of photographs?

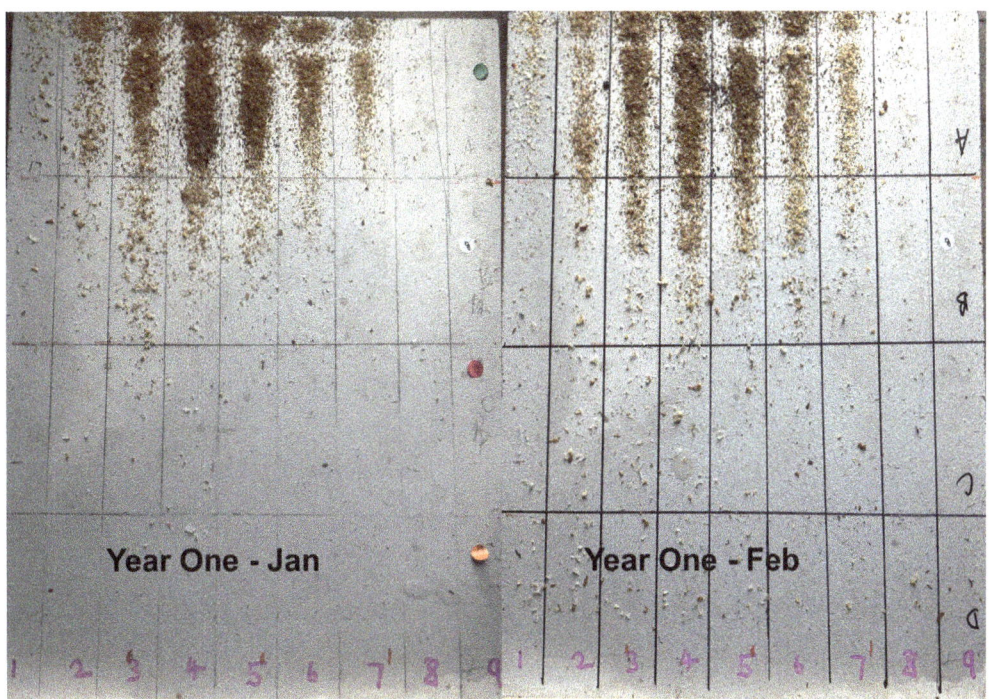

Figure 5.1 Year One Debris for January and February

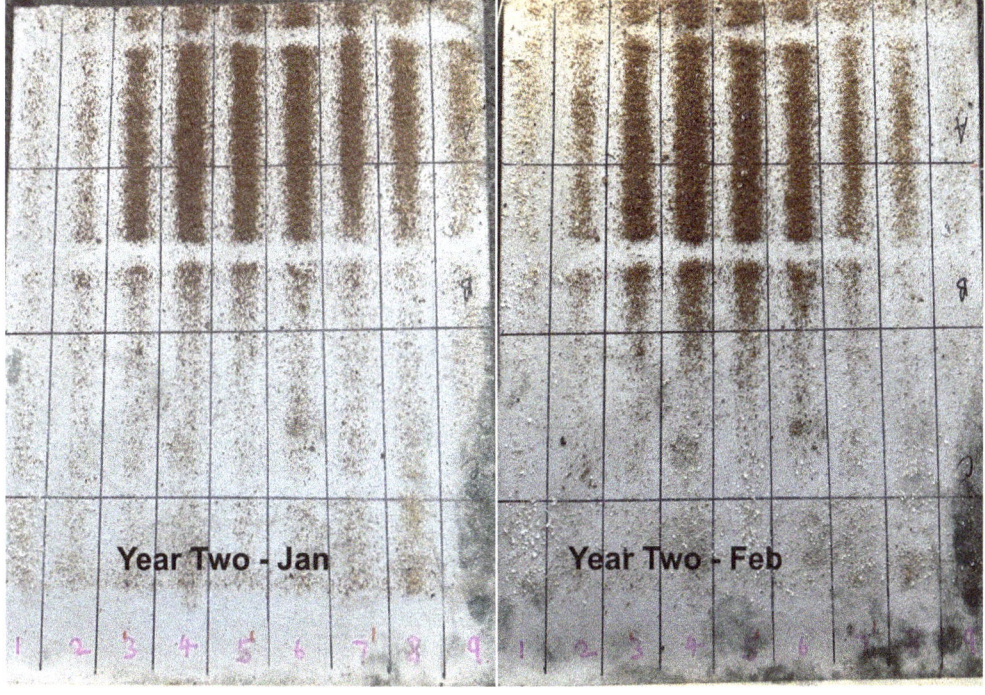

Figure 5.2 Year Two Debris for January and February

- This study has demonstrated that the debris shown by figure 5.1 came from a colony with a queen that went on to increase her rate of egg laying to produce a colony that would produce enough honey so that no supplemental feeding was needed, while also producing a healthy surplus of honey for personal consumption. A colony that also managed a significant chalkbrood burden and controlled a varroa population with minimal human intervention.
- Figure 5.2 shows debris produced by the same colony one year later. A visual comparison of these two pictures shows more debris deposition in Year Two. As a proxy indicator for colony activity it suggests that this colony is larger and more active than it was one year previous. This colony will have an increased foraging potential early in the year, which seems like a good thing. However, in poor weather a higher population could become a potential burden on existing food stores, should the weather not be favourable for flowering and foraging. It is reassuring to see that at the time of sampling that these honey bees were enjoying good weather and busy foraging on spring heather.
- In both years there was an increase in the debris deposited in February compared to January, suggesting that there had been a similar increase in colony activity. While in Year 1 the debris was diligently weighed, it soon became apparent that this measurement was not needed to make monthly comparisons. With practice it became straightforward to use the quadrats on the inspection board to make a reliable visual estimation of the trend. However, with this kind of visual estimation it is important to note that the debris has a third dimension of height. With practice it became possible to spot 'taller' debris from pictures because the higher debris peaks, the more the debris cascades to fill more of each seam, therefore reducing the white area visible on the inspection board.
- Both years shower a lighter coloured whitish debris away from the brood nest, consisting of wax of various kinds from honey bees accessing their winter stores. More of this whitish debris is visible in Year 2 compared with Year 1.

Unlike previous years, I'm no longer planning to disturb the brood nest in March. Instead, I am choosing humility over hubris and planning to spend more time, continuing monitoring debris and using the learning from Year 1 to make predictions about likely colony activity over the coming months. For example, should a similar rate of March/April expansion occur in Year 2 as witnessed in Year 1 could this colony be a contender to swarm early? I shall continue watching the debris and using insights to guide future interventions. I have adapted my National hives too, so that they collect debris and more data in the same way as described in this study. Figure 5.3 shows a sliding plywood box/inspection tray that sits under the ventilation screen. This arrangement is preferred over the traditional inspection board because the bottom and sides are completely sealed. The only way to reach the inspection board is through the ventilation screen above.

Figure 5.3 Modified National Debris Collector.

I also plan to spend more time supporting these bees from outside of the brood nest and outside of the hive. In near future, small practical steps will include slight modifications to the interior walls of the hive so that more propolis deposition is encouraged and by ensuring that the frames on the outside of the brood nest are ready for the expected brood expansion. On the outside of the hive this will involve planting more forage, in particular flowering plants such as poppies, phacelia, marigolds and feverfew may help to provide nectar and pollen to fill any June gap.

During this study I started to use the debris as a guide as to when and how I carried out hive interventions. A few examples of these interventions included giving a much more targeted treatment of oxalic acid to only the areas occupied by brood. The debris helped to pinpoint exactly where the brood was and reduce 'overspray'. In a similar way, the debris helped to narrow down where to look for eggs when I was concerned about the queen, or to check food stores, queen cells, chalkbrood infestation and so on. In all these cases, 'reading' the debris reduced the disruption to the colony. Individually, these may seem like small things, but each intervention is a potential stress factor and as discussed earlier, there is merit in working to reduce colony stress.

Imagine if I had a debris photo record for several hives over period of 10 years or more. A simple photograph of the debris taken on the 28th day of each month. I have little doubt that this would produce more reliable and valid information about how debris relates to colony activity and the wider environment. I would start to see new connections that would develop new ideas and insights that would help to make better decisions about how and when to complete management actions in the colony. That's my plan for the next ten years – perhaps you'd like to join me!

Bibliography

Adams, M. A., 2021. Pollen grains & honeydew: A guide to Identifying the plant sources in honey. 1st ed. West Yorkshire: Northern Bee Books.

Aeschlimann, M., Guangya, L., Kanji, Z. A. & Mitrano, D. M., 2022. Potential impacts of atmospheric microplastics and nanoplastics on cloud formation processes. Nature Geoscience, 15(https://doi.org/10.1038/s41561-022-01051-9), pp. 967-975.

Alma, A. M., Groot, G. S. d. & Buteler, M., 2023. Microplastics incorporated by honeybees from food are transferred to honey, wax and larvae. Environmental Pollution, 320(https://doi.org/10.1016/j.envpol.2023.121078).

Ballouk, M. A.-H. et al., 2025. Propolis mouthwashes efficacy in managing gingivitis and periodontitis: a systematic review of the latest findings. BDJ Open, 5(https://doi.org/10.1038/s41405-025-00294-z), p. 11.

BeeAware, 2024. Chalkbrood Disease. https://beeaware.org.au/archive-pest/chalkbrood/ ed. Australia: BeAware.

Bell, C. H., 2014. Food Safety Management, https://doi.org/10.1016/B978-0-12-820013-1.00063-2: USDA.

Bevan, E., 1838. The honey bee: Its natural history, physiology and management. 1st ed. London: Van Vorst.

Bloch, G., Bar-Shai, N., Cytter, Y. & Green, R., 2017. Time is honey: circadian clocks of bees and flowers and how their interactions may influence ecological communities, https://doi.org/10.1098/rstb.2016.0256: The Royal Society.

Brooke, L., 2022. The role of beekeeping management systems on microbes, pathogens, and immunity in honey bees, https://etda.libraries.psu.edu/catalog/27492bll218: Masters Thesis.

Butler, C., 1623. The feminine monarchie. West Yorkshire: Northern Bee Books.

Chapman, N., 2018. Pollen microscopy. 2nd ed. Penrith: CMI Publishing.

Cheshire, F. R., 1886. Bees & bee-keeping; scientific and practical. A complete treatise on the anatomy, physiology, floral relations, and profitable management of the hive bee. https://archive.org/details/beesbeekeepingsc02ches ed. London: The Bazaar, Exchange and Mart.

Chilcott, A., 2025. Overwintering on Honey: Is it really any better for bees than artificial feed?, https://www.beelistener.co.uk/bee-health/overwintering-on-honey-is-it-really-any-better-for-bees-than-artificial-feed/: The BeeListener.

Chilcott, A. & Seeley, T., 2018. Cold flying foragers : Honey bees in Scotland seek water in winter, Vol 158 No 1 p75: American Bee Journal.

Collison, C., 2015. A closer look : tarsal glands / footprint pheromone, Last accessed 05/03/25 https://www.beeculture.com/a-closer-look-tarsal-glands-footprint-pheromone: Bee Culture The Magazine of American Beekeeping.

Cote, D. M., Achouri, A., Karbourne, S. & L'Hocine, L., 2022. Faba bean: an untapped source of quality plant proteins and bioactives. Nutrients, 14(http://dx.doi.org/10.3390/nu14081541), p. 1541.

Dodd, M., 2000. Reading and writing the book of nature: Jan Swammerden. Endeavour, 23(https://doi.org/10.1016/S0160-9327(00)01306-5), pp. 122-128.

Edo, C. et al., 2021. Honeybees as active samplers for microplastics. Science of The Total Environment, 767(https://doi.org/10.1016/j.scitotenv.2020.144481).

Ghosh, S., Jeon, H. & Jung, C., 2020. Foraging behaviour and preference of pollen sources by honey bee (Apis mellifera) relative to protein contents. Journal of Ecology and Environment, 1(http://dx.doi.org/10.1186/s41610-020-0149-9), p. 44.

Han, B., Wu, J. & Wei, Q., 2024. Life-history stage determines the diet of ectoparasitic mites on their honey bee hosts.. Nature Communications, 15(https://doi.org/10.1038/s41467-024-44915-x), p. 725.

Hodges, C. R. L., Delaplane, K. S. & Brosi, B. J., 2019. Textured hive interiors increase honey bee (Hymenoptera: Apidae) propolis-hoarding behavior. J Con Entomol, 2(https://doi.org/10.1093/jee/toy363), pp. 986-990.

Kahn, K. A. & Liu, T., 2022. Morphological structure and distribution of hairiness on different body parts of apis mellifera with an Implication on pollination biology and a novel method to measure the hair length. Insects, 13(https://doi.org/10.3390/insects13020189), p. 2.

Kubasek, J., Svobodova, K., Puta, F. & Krejci, A. B., 2022. Honeybees control the gas permeability of brood and honey cappings. iScience, 11(https://www.sciencedirect.com/science/article/pii/S2589004222017175), p. 25.

Kwong, W. & Moran, N., 2016. Gut microbial communities of social bees. Nat Rev Microbiol, 14(https://doi.org/10.1038/nrmicro.2016.43), pp. 374-84.

Langstroth, L. L., 1880. A practical treatsie on the hive and the honey bee. 3rd ed. New York: C M Saxton, Barker & Co.

Lawrence, D. P., Rotondo, F. & Gannibal, F. B., 2016. Biodiversity and taxonomy of the pleomorphic genus Alternaria. Mycology Progress, 3(https://doi.org/10.1007/s11557-015-1144-x), p. 15.

Leponiemi, M. et al., 2023. Honeybees' foraging choices for nectar and pollen revealed by DNA metabarcoding. Science Report, 13(https://doi.org/10.1038/s41598-023-42102-4), p. 7.

Lopez-Uribe, M. M. & Lawrence, B., 2021. The biology of the bread that bees make: microbes, biocide, and fermentation in the hive, https://www.researchgate.net/publication/356770554_The_Biology_of_the_Bread_that_Bees_Make_Microbes_Biocide_and_Fermentation_in_the_Hive: Fermentology.

National Bee Unit Website, Last accessed 05/03/25. Honey bee starvation, https://www.nationalbeeunit.com/about-us/beekeeping-news/recognise-the-signs-of-starvation.: Bee Base.

National Bee Unit, 2025. Spring checks, https://www.nationalbeeunit.com/assets/PDFs/3_Resources_for_beekeepers/Best_practice_guidelines/BPG_6_Spring_Checks_2018.pdf: Animal & Plant Health Agency.

Nguyen, T.-T.et al., 2024. Carpoglyphus ;actis (acarina: carpoglyphidae) in apis mellifera in the Republic of Korea. Insects, 15(https://doi.org/10.3390/insects15040271), p. 271.

Nurnberger, F., Hartel, S. & Steffan-Dewenter, I., 2018. The influence of temperature and photoperiod on the timing of brood onset in hibernating honey bee colonies. PeerJ Life and Environment, Issue https://doi.org/10.7717/peerj.4801.

Palgrave, C., 2022. Honey bee health & welfare, https://www.beelistener.co.uk/bee-health/honey-bee-health-welfare-dr-chris-palgrave-comments/: The Bee Listener.

Payne, Bonwen; Campbell, Lucy; Robertson, Jessica; Marriot, Sara; MacKinnon, Cael; Scott, Aaron; Redpath, Jamie; Yekken, Zinedine; Ferramenta, Mateus Gomes; Todd, Demi; Millburn, Ella; North, Israel; 2024 Supporting teachers - Angella Yekken, Lesley Rosher, Jen Addie & Ray Baxter. Scottish schools study into microplastics found in honey bee colonies. Berwickshire High School, Kelso High School & St Ninians High School.

Pestium, 2024. The prune mite, https://www.pestium.uk/food-pests/the-various-species/the-prune-mite/.: Website last accessed 03/03/25.

Pirk, C. W., Neumann, P., Hepburn, R. & Tautz, J., 2004. Egg viability and worker policing in honey bees. PNAS, 101(https://doi.org/10.1073/pnas.0402506101), p. 23.

Ponting, S. & Stainton, K., 2020. Chalkbrood. BBKA News - Incorporating the British Bee Journal, Issue https://www.nationalbeeunit.com/assets/PDFs/3_Resources_for_beekeepers/articles_reports/BBKA_news/BBKA_61_Chalkbrood_July_2020.pdf, p. 237.

Poole, J., Church, J. S., Wilde, A. & Huson, M., 2013. Continuous production of flexible fibers from transgenically-produced honeybee silk proteins. Macromolecular Bioscience, 10(http://dx.doi.org/10.1002/mabi.201370033), p. 13.

Posada-Florez, F. et al., 2021. Pupal cannibalism by worker honey bees contributes to the spread of deformed wing virus. Nature, 11(https://doi.org/10.1038/s41598-021-88649-y), p. 8989.

Pusceddu, M. et al., 2021. Honeybees use propolis as a natural pesticide against their major ectoparasite, https://doi.org/10.1098/rspb.2021.2101: The Royal Society.

Pusceddu, M. et al., 2019. Resin foraging dynamics in varroa destructor-infested hives: a case of medication of kin?. Insect Science, 26(https://doi.org/10.1111/1744-7917.12515), pp. 297-310.

Rosenkranz, P., 2020. Biology and control of varroa destructor. Invertebrate Pathology, 1(doi: 10.1016/j.jip.2009.07.016), pp. 96-119.

Rutkowski, D., Weston, M. & Vanette, R. L., 2023. Bees just wanna have fungi: a review of bee associations with nonpathogenic fungi. FEMS Microbiology Ecology, 99(https://doi.org/10.1093/femsec/fiad077).

Santomauro, G., Oldham, N. J., Boland, W. & Engels, W., 2004. Cannibalism of diploid drone larvae in the honey bee (Apis mellifera) is released by odd pattern of cuticular substances. Journal of Apicultural Research, 2(http://dx.doi.org/10.1080/00218839.2004.11101114), p. 43.

Saunders, D. S., 2012. Insect photoperiodism: seeing the light, https://doi.org/10.1111/j.1365-3032.2012.00837.x: Royal Entomological Society.

Saunders, D. S., 2012. Insect photoperiodism: seeing the light. Physiological Entomology, 37(https://doi.org/10.1111/j.1365-3032.2012.00837.x), pp. 207-218.

Sawyer, 1981. Pollen identification for beekeepers. 1st ed. Cardiff: University College Cardiff Press.

Shanahan, M. et al., 2024. Thinking inside the box: Restoring the propolis envelope facilitates honey bee social immunity. PLOS ONE, Issue https://doi.org/10.1371/journal.pone.0291744.

Siefert, P., N, B. & B, G., 2021. Honey bee behaviours within the hive: Insights from long-term video analysis. PLOS One, 3(https://doi.org/10.1371/journal.pone.0247323), p. 16.

Simmons, E. G., 2007. Alternaria An Identification Manual. s.l.:Centraalbureau voor Schimmelcultures.

Steen, J. v. d., 2016. Behold : the colony of the honey bee as a biosampler for pollutants and plant pathogens, https://research.wur.nl/en/publications/beehold-the-colony-of-the-honeybee-apis-mellifera-l-as-a-bio-samp: Entomology & Disease Management, Environmental Technology, WIMEK.

Stelluti, F., 1630. Persio tradotto in verso sciolto e dichiarato da Francesco Stelluti …. 1st ed. Rome: Appresso Giacomo Mascardi.

Sumner, S., 2022. Endless forms. 1st ed. Glasgow: William Collins.

Sutherland, T. D. et al., 2006. A highly divergent gene cluster in honey bees encodes a novel silk family. Genome Res, 11(https://doi.org/10.1101/gr.5052606), pp. 1414-21.

Sutherland, T. D. et al., 2011. Single honeybee silk protein mimics properties of multi-protein silk. PLoS One, 2(https://doi.org/10.1371/journal.pone.0016489), p. 6.

The National Bee Unit, 2024. Varroa best practice guidelines, York: Animal & Plant Health Agency.

Traynor, K. S., Mondent, F., Miranda, J. R. & Maeva Techer, 2020. Varroa destructor: a complex parasite, crippling honey bees worldwide. Trends Parasitol, 36(7)(https://doi.org/10.1016/j.pt.2020.04.004), pp. 592-606.

USDA, 2012. Report on the national stake holders conference on honey bee health, https://www.usda.gov/sites/default/files/documents/ReportHoneyBeeHealth.pdf: USDA.

USDA, 2016. Bee Mite ID: Bee-associated Mite Genera of the World. Website: https://idtools.org/bee_mite/.

Visscher, P., 1989. A quantitative study of worker reproduction in honey bee colonies. Behavioural Ecology and Sociobiology, 25(https://doi.org/10.1007/BF00300050), pp. 247-254.

Wilmart, O. et al., 2021. Honey bee exposure scenarios to selected residues through contaminated beeswax. Sci Total Environ, Issue https://doi.org/10.1016/j.scitotenv.2021.145533, p. 772.

Bottom-Up Beekeeping *Ray Baxter*